GETTING STARTED IN SCALE MODELING

U.S. EDITION

An AK-Interactive Book

Kalmbach Media
20127 Crossroads Circle
Waukesha, Wisconsin 53186
www.KalmbachHobbyStore.com

First published in the United States in 2019 for exclusive North American distribution by Kalmbach Media.

Published in 2019
23 22 21 20 19   1 2 3 4 5

Manufactured in China

ISBN: 9781627007146
EISBN: 9781627007153

Conceived, designed, and produced by
AK-Interactive, S.L.
C/ Del Chozo 6
26006 Pol. Ind. La Portalanda Logrono
Spain

www.ak-Interactive.com

Originally published in Spain under the title "GUÍA DE MODELISMO PARA PRINCIPIANTES."
Original idea and concept: Eduardo Fernández Rodríguez, AK-Interactive.
Project manager: Eduardo Fernández Rodríguez.
Artistic direction: AK-Interactive.
Graphics & layout: BMS Designs, AK-Interactive.

Project manager, U.S. edition: Eric White
Cover design, U.S. edition: Lisa Bergman
English translation edited for U.S. edition by Tim Kidwell

Library of Congress Control Number: 2019935739

Produced & distributed by:

www.ak-interactive.com   Follow us on 

Index

3. PAINTING

4. PAINTING TECHNIQUES

Introduction

Scale modeling. Our hobby!

Scale modelers typically get into the hobby because they've seen beatuifully built and finished models in stores, on the internet, or in movies. They think, I can do that too!

Modeling is an enormously satisfying hobby, accessible to anyone willing to work on their skills. But that's the key: skills. Often, when a first-time modeler buys that first kit, no matter the subject, the stack of unpainted plastic parts inside the colorful package can be daunting. They might even think the task too big, and abandon hope of assembling the kit.

First-time modelers face a seemingly endless number of questions: Which tools do I need to build a model? What kind of paints should I use? I saw a really cool effect on another model, so how do I replicate it? Where do I start?

There are many resources that explain how to assemble model kits. But when you're new to the hobby, many of those resources can create more questions and confusion because they target more experienced modelers, taking basic modeling knowledge for granted.

With this book, we want to help novice modelers understand basic modeling tools and skills. Building scale models isn't difficult. Rather, it's through the consistent applications of these basic skills that you advance your abilities and achieve satisfying results.

In the following chapters, we detail all the essential techniques for building scale models. You'll learn when and how to apply them. We also review basic painting techniques, and explore a variety of finishing alternatives. Everything you'll learn in this guide can be applied to almost any plastic model, regardless of scale or subject. Lastly, we hope this guide will eliminate any doubts you may harbor and help you to confidently immerse yourself in the wonderful world of scale modeling.

1. First steps and basic terms.

Choose wisely before you start

How often have we heard you that you need to learn to walk before you can run? It may be frustrating, but it also applies to scale models. If you're new to scale modeling, the best thing you can do is choose a simple kit with easy construction, clear instructions, and parts that fit well together. Nowadays, most plastic kit makers produce a range of kits that are beginner-friendly, and have all these aspects resolved to a high degree. Choosing a simpler kit can limit the range of subjects available to you early on. However, diving into a kit with resin and photo-etched parts can prove overwhelming and stifle your progress or convince you to quit altogether. Better to start with a model you're likely to finish.

Model kits come in a variety of materials, including styrene plastic, resin, brass, white metal, and even wood. Each of these materials poses its own challenges when it comes to assembly and finishing. Plastic kits far and away dominate the scale-model market, and are less demanding in terms of tools and equipment. While many of the more complex model kits can be multimedia (meaning they include parts made from several materials), at first, we recommend you start with all-plastic kits.

And speaking of kit complexity, the market is full of model kits advertising a specific skill level. While there is no consistent rating system, these numbers can be useful. At the least complex end of the range the so-called "easy-kits." Sometimes these come prepainted or are made with plastic that doesn't require paint. These kits are perfect for children because they offer a quick result, easy entry into the hobby, and sometimes are rugged enough to play with. Simpler plastic kits in the 1 or 2 skill range can provide a quick and smooth entry into modeling to teens and adult novices. When you see kits that require skill levels of 3 or higher, you know you're getting into territory beyond a beginner's abilities. There is no shame associated with purchasing and building an easy kit. You want to succeed, so sticking with easy kits at first makes the best sense. Soon, your skills will improve and you'll be building more complicated projects in no time.

To a scale modeler, a hobby shop is like a huge amusement park. With the plethora of models available, it's easy to get lost—we want everything! But to be able to succeed and advance your skills, you have to start with something simple that motivates you. With time and experience, the other kits that catch your eye will be well within your abilities.

A scale model's material often dictates the types of tools, paint, and finishing products you'll use during assembly.

PLASTIC

The majority of commercial kits are made of injected plastic. Parts are numbered for easy identification and come attached to a frame called a sprue. Following the kit instructions, you remove the pieces from the sprue and smooth out any irregularities. Typically, you use plastic cement (which is often a gel), liquid cement, or super glue. Depending on the quality of the parts fit, sometimes joints and gaps must be filled with modeling putty and sanded for a smooth finish.

RESIN

Polyurethane resin is often used for figures, accessories, and upgrades. Resin parts usually come molded on resin blocks, called runners. Resin parts require a saw or hobby knife with a stout blade and handle to remove the parts from the runners because of the density and hardness of the material. Resin parts can be assembled using either super glue or two-part epoxy. Resin parts should be washed and primed before any kind of paint is applied. **Important:** Resin dust can be extremely harmful when inhaled. Always use a mask and gloves when sawing or sanding resin parts.

CARDBOARD AND PAPER

While kits made of cardboard or paper are rare, the use of these rather inexpensive materials is steadily gaining popularity. Cardboard and paper parts usually come in pre-cut sheets. White glue or super glue can be used for assembly. The cardboard and paper pieces must preferably be sealed with varnish before painting.

WOOD

You usually see wood used for ship model kits and radio-control airplanes. Wooden parts are often found packaged in plastic bags or attached to laser-cut sheets similar to cardboard and paper kits. Depending on the subject, the type of wood ranges from thin plywood to basswood to balsa. You'll need a sharp hobby knife or razor saw to cut wood parts. Wood parts should always be sanded smooth. Finishing wooden models can include primer and paint. However, some decide to use stains or oil finishes for a natural wood look. Wood glue, white glue, and super glue all work on wooden models.

METAL

It's rare to encounter a scale model made completely from metal. However, white metal is frequently used for figures and upgrade parts for armored vehicles, airplanes, and ships. The most common metal parts are photo-etched brass. Before painting, metal parts should be thoroughly cleaned and primed. Super glue and epoxy can be used for assembly.

POLYSTYRENE FOAM AND FOAM CORE

Modelers use polystyrene foam and foam core for scratchbuilding, and it isn't typically found in commercial model kits. Polystyrene foam is useful for making large structures and forming landscapes or terrain. Polystyrene foam can be found at hardware and home stores in the insulation section; foam core is generally available in the craft aisle at your local store.

Choosing the subject, scale, and materials

Choosing a model to build need not be difficult. Keep a few parameters in mind: First, what are you interested in? Do you like figures, tanks, aircraft, cars, ships? Is fantasy or science fiction more your speed, or do you prefer models of true-to-life subjects? It's important to pick a subject that inspires you, because that will hold your interest all the way to the end of the build.

Second, what scale are you interested in building? Of course, you have to understand what scale is before you can make that decision.

The objective of scale modeling is to build a replica of a subject in miniature. Scale is the numerical relation between the larger subject and its reduced size and is typically noted as a fraction or ratio. For example, a replica built at full size would be 1/1 (1:1) scale, meaning 1" on the replica would equal 1" on the subject. If you are building a 1/4 scale model, 1" on the model would represent 4" on the full-size subject, making the model one-quarter the actual size. Put another way, if you made a 1/4 scale model of a canoe measuring 12 feet long, the model would be 3 feet long.

The larger the denominator in the scale fraction, the smaller the scale. For example, a 1/72 scale helicopter model is 72 times smaller than it's full-size counterpart. A 1/32 scale model of the same helicopter is 32 times smaller than the original, and would therefore be larger than the 1/72 scale model. The scale of a kit is usually displayed prominently on the box.

The size of your workspace, the steadiness of your hands, and even your eyesight can affect your choice of scale. Make sure to take into account these considerations as well as your display options when the model is finished when choosing your kit's scale.

Sometimes, usually when referring to figures or wargame vehicles, you'll see scale expressed in millimeters. For example, a figure can might be labelled 54mm or 75mm. This proportion refers to the size of the figure, measured from the feet to the head. Assuming the height of a man depicted at 54mm is 6 feet (1,829mm) tall, dividing the scale by the full height will give you a sense of the traditional scale expression. For example, a 54mm figure, is roughly equivalent to 1:35 scale.

When building a single model, the only thing you only have to consider is the scale of the model you want to build. However, if you want to depict a scene combining many different elements, lets say a figure, a car, and a building, you have to make sure that all the elements are in scale. If you don't, the result will look strange—what's known as out-of-scale.

Scale figure too small.

Scale figure too large.

Correct scale.

The last thing you need to consider before deciding on a model kit is its material:

PLASTIC KITS

These kits are the most common on the market. Depending on the manufacturer and pricing, they incorporate a high level of quality, detail, and accuracy when compared to the full-size subject. They are made by injecting colored plastic into metal molds.

METAL KITS

Metal kits are most often figures or upgrades and aftermarket parts for commercial kits. Originally, metal kits were manufactured by pouring lead into a mold, but this metal has been replaced by other "white metal" alloys. Photo-etched brass parts are also found in this category.

RESIN KITS

Resin kits typically cost more than plastic kits due to limited supply and high production costs. Quality varies widely with resin kits, so it's always a good idea to have a careful look inside the box before purchasing one.

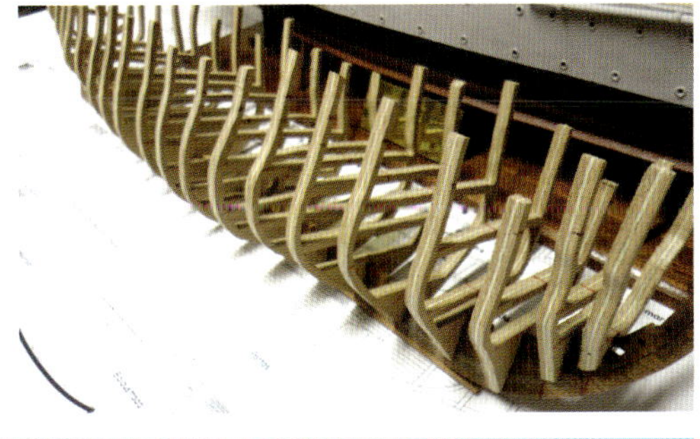

WOODEN KITS

Used almost exclusively for ship models, wooden kits require a number of special tools for building and finishing, and tend to be complicated affairs.

TIP

- Check proportions when you compose different elements in a scene. Always know the scale you're building in and keep a calculator handy.

Preparing your workspace

Before starting your first model, you should establish a dedicated workspace—your workshop. Building models can sometimes become dirty, untidy, and noisy. You probably don't want the mess you can create out in the open. Finding a spot out of the way and relatively private is probably the best location for your workbench. The choices you've made concerning the subject, scale, and material composition of the models you will build will affect the amount of space you'll need. As you would expect, a workshop dedicated to figure painting can be considerably smaller than one intended for building wooden ships. In any case, your new workshop has to meet some basic requirements in order to be safe, functional, and comfortable.

THE WORKBENCH

The central focus for any workshop is its workbench. To build your models, you will need a stable, level, horizontal surface large enough to host the average size of the type of kits you've decided to build. It can be as simple as an old table or desk, or might be a specially built workbench. Whatever it might be, remember that the tools, paints, solvents, and other materials you'll use can cause damage. A thick piece of glass or a self-healing cutting mat can be helpful to reduce potential damage. Cutting mats are generally affordable, and can be found in different sizes and colors. Many come with a printed grid, which can be useful for quick measurements.

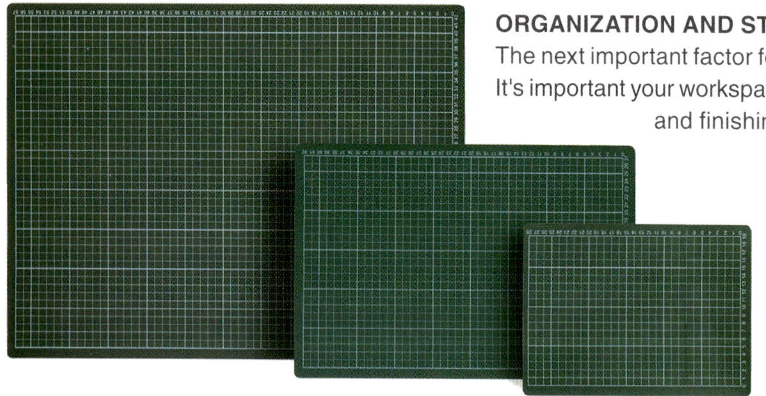

ORGANIZATION AND STORAGE

The next important factor for a functional workshop is organization and storage. It's important your workspace remains tidy and organized during the construction and finishing stages of a build.

Considering the number of tools and materials you'll use during the different phases of construction, disorder can easily rear its ugly head after only a few minutes of work. To avoid this, cutters, sanders, tweezers, paints and solvents—everything—should have its own space in your workshop.

Thankfully, a vast range of organizers are available. Designed to rest on your workbench or hang on a wall or backboard, these purpose-made organizers can help you keep your tools organized and accessible at all times.

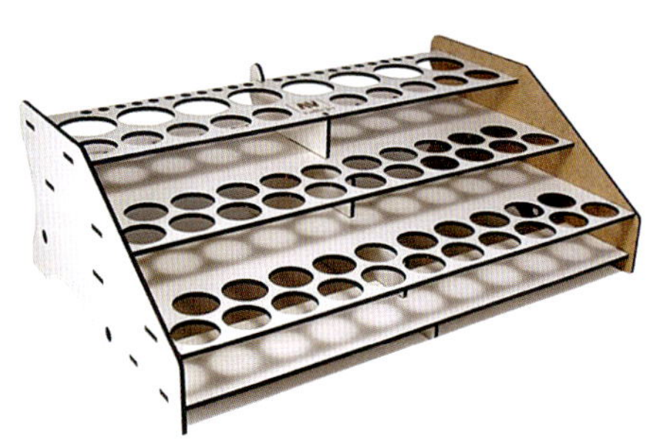

For bulkier tools, such as airbrush compressors and motor tools, specific space must be defined, keeping in mind that all regulating parts (pressure valves, selectors, gauges, pedals, etc.) must be within sight and easy reach.

LIGHTING

You should have dedicated light sources for your workspace. There's no need to get fancy—almost any desk or office lamp can provide the light you need. What's more important is quality of light. White light is the ideal light for modeling. Warm light tends to be yellow and makes colors appear dark or dirty. Also, consider where you place your lights. You want to avoid creating shadows and blind spots. For example, a right-handed modeler should place the main light source on the left, and vice versa for a left-handed modeler. To minimize the danger of inadequate lighting, consider using multiple light sources. This is easier to do when you have a dedicated workspace where you can set up permanent fixtures.

A poorly placed light will cause your hand to project shadows over your work, hinder visibility, and lead to eye fatigue.

The good placement of your light source eliminates unwanted shadows, brings out details, and reduces eye strain.

COMFORT

Scale modeling is a static and time-consuming hobby. Building and painting sessions can last hours. Thus, you'll want a comfortable, adjustable chair for your workshop. Office chairs can be found in a wide range of prices and quality. Use a chair that matches your physical features and needs. Long hours in the seated position can cause strain to your back and neck, so consider a chair that promotes good posture. Physicians suggest avoiding sitting in the edge of the seat, while our back should rest evenly on the chair's backrest. Also, be sure to get up periodically to stretch and walk around.

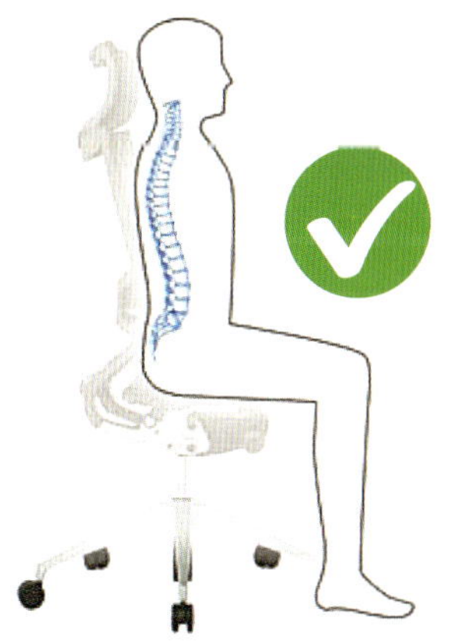

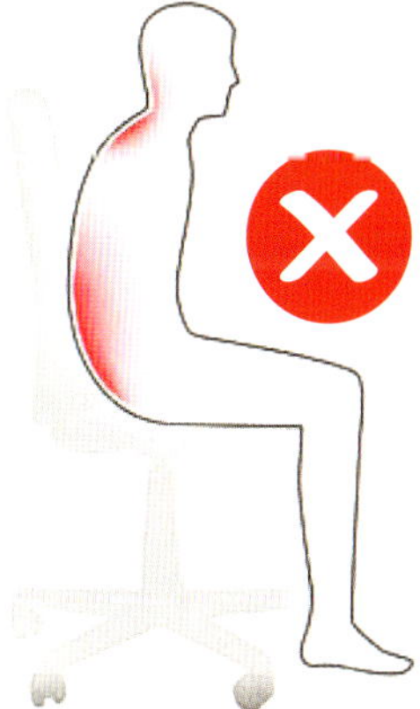

TIPS

- Keep a container of water and a clean cloth or paper towels close at hand to clean the workbench when needed.
- Your workspace should be well ventilated; many modeling products emit vapors that at best have a strong odor or at worst can be toxic or flammable.
- If you store your paint in drawers or boxes with the caps up, put a drop of paint on the cap for quick identification.

Cutting tools

When looking at an experienced modeler's workbench, the number and variety of tools on hand often amazes new modelers. Don't let the wide array of tool choices available discourage you—you don't need them all, and you'll have plenty of time to acquire them over the course of your years in the hobby. All a new modeler needs to get building are a few basic tools.

Cutting tools are the most basic and essential tools for a scale modeler. They're used for cutting, trimming, scratching, scribing, carving, and much more. Cutting tools come in many varieties and all have their uses—some more specific than others. The tools listed here will meet most any modeling challenge you'll face.

HOBBY KNIFE

The so-called scalpel is the iconic scale modeling tool. Derived from the equivalent surgical instrument, a hobby knife is an extremely sharp tool, offering precision and durability. The knife handle comes equipped with a chuck to accept a number of interchangeable blade shapes, from the standard No. 11 blade to more specialized chisel and carving blades.

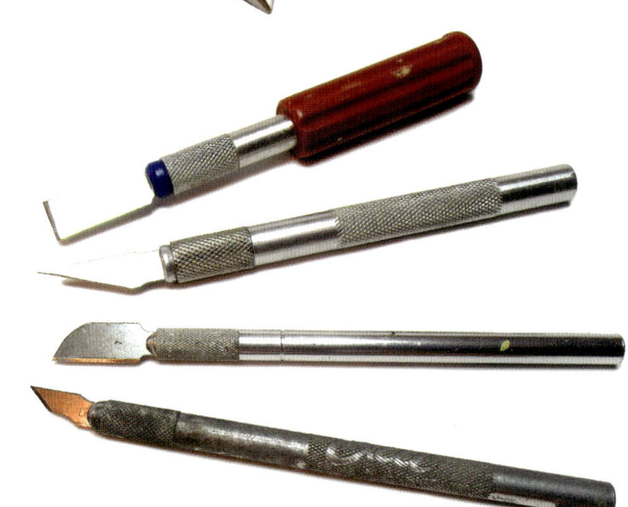

UTILITY KNIFE

A utility knife or snap-blade cutter comes in handy when you're cutting large or thick materials and you don't need a lot of precision. Most commonly you'll use it while making diorama elements for buildings or cutting materials such as heavy paper, cardboard, balsa wood, and polystyrene foam.

CUTTING PLIERS

This is another essential tool for a scale modeling workshop. They can be used to cut through rods and strips of plastic and metal. Commonly, you'll use cutting pliers, also called sprue cutters, to remove kit parts from the sprues. In their basic form they're not made for precision cutting. However, flush-cutting micro shears come with fine, sharp edges and specially shaped heads to get into narrow or difficult-to-reach areas. Not all cutting pliers can be used on metal parts, and don't confuse cutting pliers with standard pliers. They're used for very different purposes.

RAZOR SAW

A razor saw can make large cuts with considerable precision. They're useful for building model structures, and flush cuts on wooden dowels or plastic rods. Combined with a miter box, a razor saw allows you to make precision cuts for joints at a variety of angles. Razor saw come in a variety of widths and tooth counts specific for use with wood, plastic, or metal.

MOTOR TOOL

Controlling and mastering a motor tool requires patience and experience. However, they're extremely useful. A motor tool can be equipped with a wide variety of bits for routing, drilling, cutting, and sanding.

TIPS

- A hobby knife and a quality set of cutting pliers are the most essential tools for a new modeler.
- While using any modeling tool, always consider safety. Most modeling tools are sharp and can cause serious injury.
- Investing in a few quality tools is better than getting many mediocre ones. Spend wisely.

Sanding tools

Sanding tools are indispensable for any but the most basic modeling projects. Sandpaper is nothing more than a flexible medium, be it plastic sheet, paper, or cloth, covered with abrasive material—typically aluminum oxide or silicon carbide.

Sanding sticks come in many shapes and sizes. The most useful to the beginning modeler are easy to hold and control, and are flexible. You'll want both a range of sandpaper and sanding sticks in your modeling workshop.

The coarseness of sandpaper is measured by the number of the grit size. The higher the number, the finer the grit. The grit size used for scale modeling typically ranges between 200 (coarse) and 2000 (fine) and are used for a multitude of tasks.

One such sanding task is to prepare a surface that's been filled with putty for a primer coat of paint. After your model has been assembled and all the gaps filled with putty are completely dry, it's time to sand for that perfect finish.

First, remove the excess putty with 200-grit sandpaper. Focus on the filled spot to prevent damage to your piece or removing nearby detail.

Next, using 600-grit sandpaper and continuing to focus on the puttied area, gently sand until all grooves and marks caused by the previous 200-grit sandpaper have been removed, but still leaving putty in the gaps.

Move on to 1200-grit and finer sandpaper to give your surface a progressively smoother finish. Check the result by running your finger over the sanded surface. If your model requires an extremely smooth surface (a car for example), polish the area by rubbing it with a piece of parchment paper or polishing cloth.

Afterward, brush denatured alcohol onto your model to eliminate any grease or dust that has accumulated during sanding.

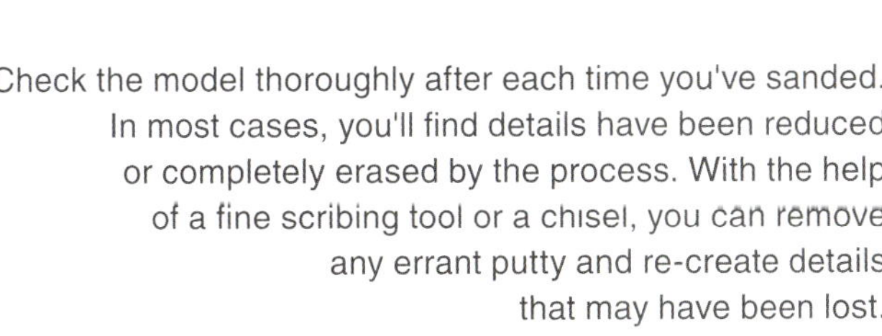

Check the model thoroughly after each time you've sanded. In most cases, you'll find details have been reduced or completely erased by the process. With the help of a fine scribing tool or a chisel, you can remove any errant putty and re-create details that may have been lost.

Sanding can be tedious, and some modelers tend to rush the process—or even omit it! Keep in mind a badly sanded model will result in a rough, uneven finish and spoil the overall quality of your project. Sandpaper is easy to acquire, so there's little excuse not to have several sheets of different grits in your workshop

TIPS

- Sanding sponges and sanding sticks are ideal for coarse sanding. For fine sanding, it's better to use sandpaper because it allows better control of the pressure.
- Dipping sandpaper in water while sanding provides smoother results, reduces scratches to nearby areas (especially when working with fine grits), and keeps the dust down.
- Sequential sanding can also be used to polish and leave paint like a mirror. Start with 1500-grit sandpaper and progress through finer grits until you reach 12000-grit.

Files and steel wool

During the construction of a model kit, you'll almost surely need to remove excess material from parts, either to correct irregularities and flaws from the production process or to improve part fit. This is job for a file or a rasp.

For our purposes, files are made of case-hardened steel, often incorporating a wooden, plastic, or metal handle.

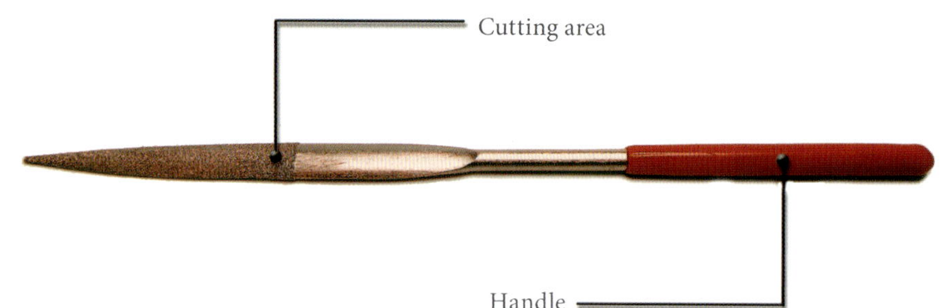

Rasps

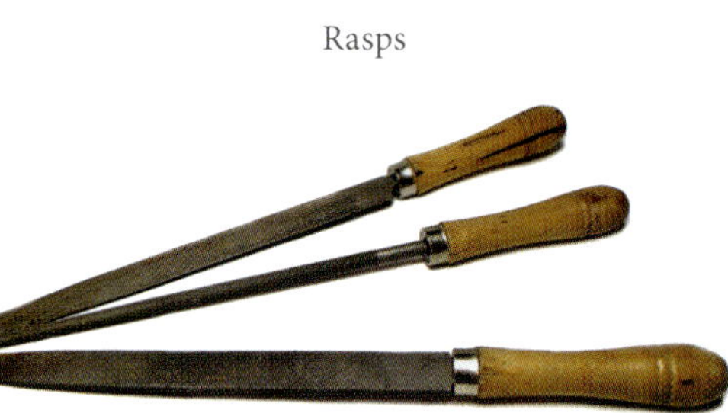

Not all files are the same. The grinding area of one file may be made from different materials than another file. Files come in a range of shapes, cross-sections, and cuts. The cut or how fine a file's teeth are give it a specific purpose and use.

Diamond files

Like sandpaper grits, files have cut grades: rough, middle, bastard, second cut, smooth, and dead smooth. The coarsest of files, commonly called rasps, are primarily used for shaping wood and removing a lot of material quickly. In contrast, diamond files (so called because of the industrial diamonds embedded in the surface) are suitable for use with very hard materials like glass and metals. Needle files, which is what modelers primarily use, are small, come in a range of shapes and cross sections, and are designed more with surface finish in mind rather than rates of removal.

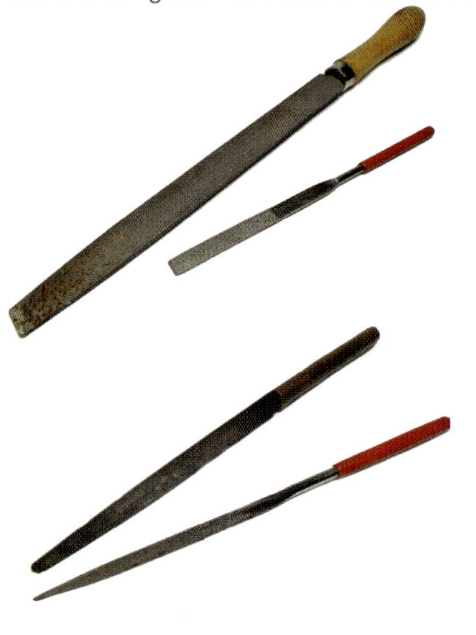

FLAT FILES
Used to level, reduce, or smooth flat surfaces that need to remain perfectly flat. They're also very useful for creating sharp angles and edges.

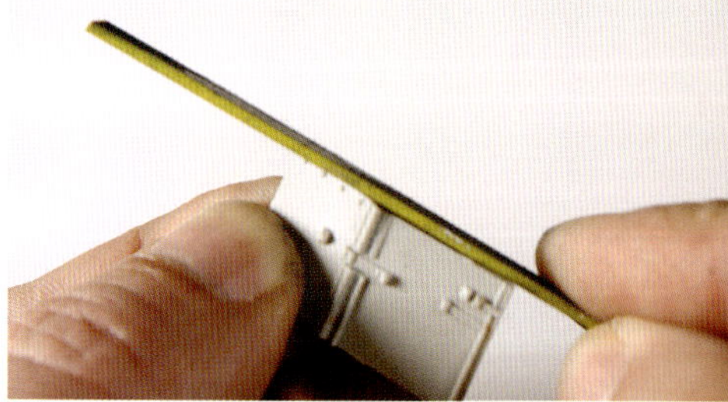

TRIANGULAR FILES
Used to simultaneously trim or shape inverted angles. They can also provide access in angled areas, create or shape recesses, and even enhance scribed or etched detail.

ROUND FILES
Used mostly to trim, reduce, or shape curved and circular areas.

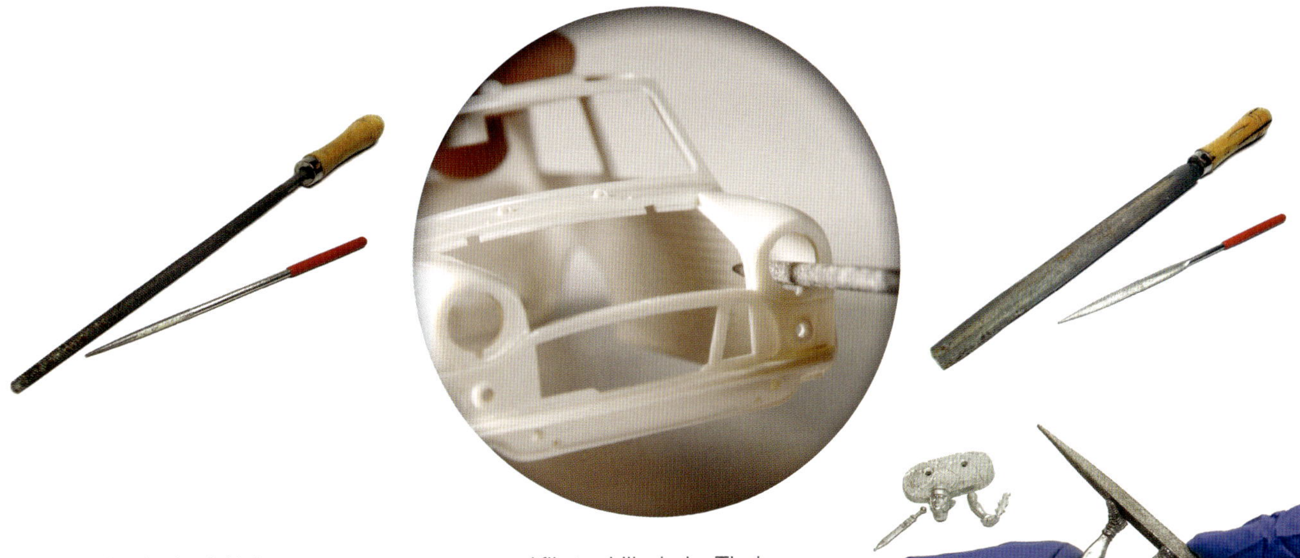

If found in the desired thickness, you can use a round file to drill a hole. Their use needs some caution and dexterity, as they don't always come in a perfect, circular shape. You can also use a half-round file to work and shape curves.

A set of files is necessary and useful, but always take care to consider the file's cut and shape so you don't remove more material than you intend.

STEEL WOOL
Steel wool (also known as wire wool or wire sponge) is a dense mass of metallic threads. It's an excellent polisher, especially on wood and metal. It can work wonders on white metal figures or kit parts. Avoid using steel wool on fine details because it might get tangled in them and cause them to break.

TIP

- Be patient when using a file. Work slowly and continuously check your progress until you're satisfied with the result.

Auxiliary tools

Modelers accumulate many tools over time. As we've seen, some of them are absolutely essential, but we acquire others because they facilitate a particular task, allowing us to advance our techniques and improve our results. In this section, we'll present the most common auxiliary tools. While no less essential than the basic tools, the auxiliaries have limited functionality and are usually employed less often.

CLAMPS

Clamps can be of great help, both during construction, and in the painting process. Thanks to their spring action, they offer a constant and secure hold of two or more surfaces, thus enabling strong bonds between parts while gluing. They can also be used as holders during painting, especially when parts need to be painted separately. Some clamps have too strong a grip and can easily bend or even break fragile parts. Make sure to check the grip of a clamp before using it.

TWEEZERS

Tweezers can be a priceless ally when working with small parts and details. You can find them in many different sizes and types, the most common being straight, angled, and flat-head tweezers. They can all grab and hold pieces more easily than your fingers. Reverse tweezers offer a self-locking grip. Overall, tweezers improve your ability to place parts in difficult-to-reach places or position parts too small for you to comfortably handle with your fingers.

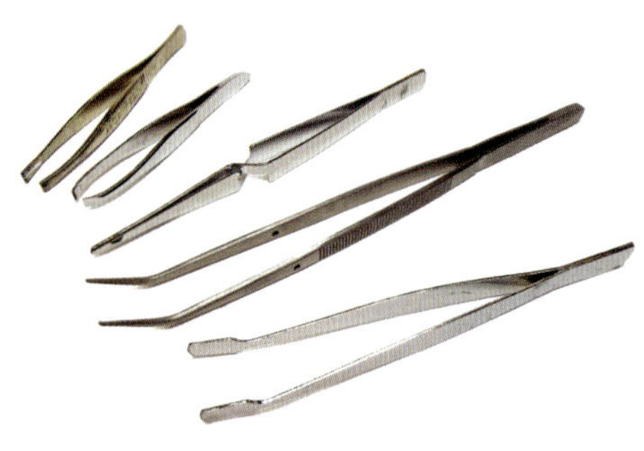

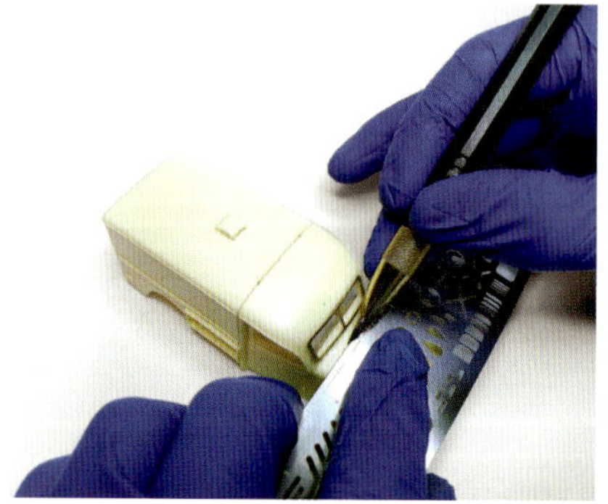

SCRIBERS, CHISELS, RIVET MAKERS

Scribers are used during construction to enhance or create new scribed or etched detail, such as panel lines and recesses that are lost because of sanding and filling, or must be added because the details simply didn't exist on the kit parts.

Rivet makers are scribers with a notched rotating head that leaves lines of evenly spaced marks as you roll it along a surface. Rivet makers, also called pounce wheels, are most commonly used on aircraft kits to add or restore damaged detail. Chisels are sculpting tools for removing large amounts of material, but you can employ them as scribers in a pinch. Using differently shaped and sized chisel heads, you can easily create whatever recesses and grooves you may require.

DROPPERS, PIPETTES

Droppers and pipettes may look very scientific, but they become extremely handy when painting. Look for droppers or pipettes that are graduated (usually ounces or milliliters) so you can accurately measure the amounts you use them to transfer. Pipettes are often made of plastic and disposable. Droppers can be made from plastic or glass, but are meant to be cleaned and reused.

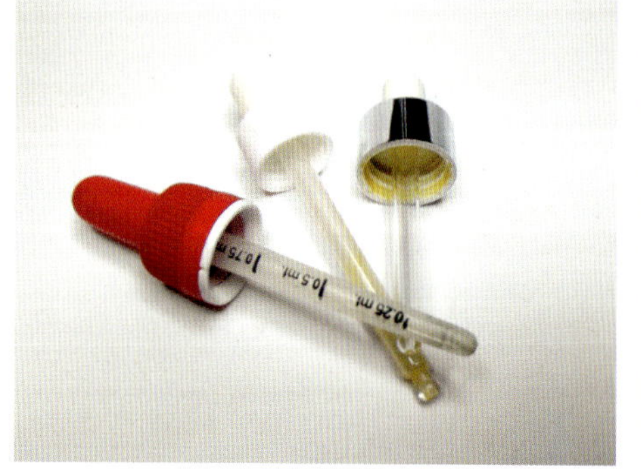

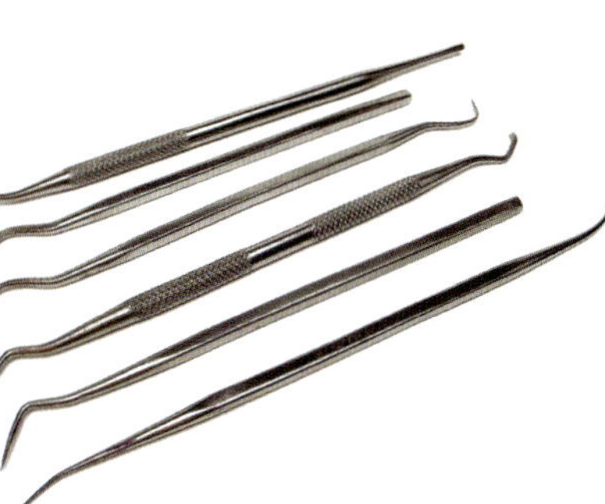

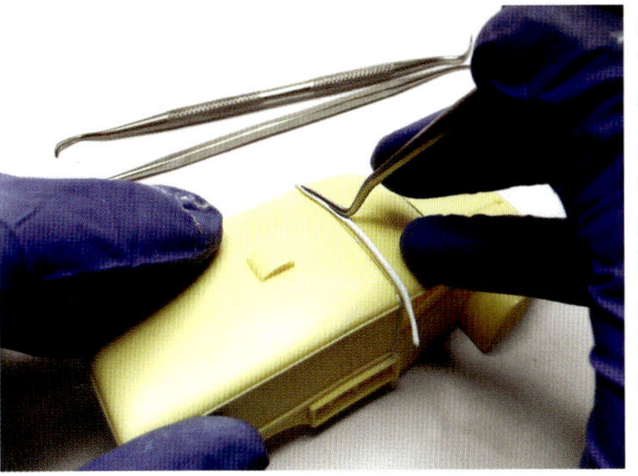

SCULPTING TOOLS

These tools derive mainly from dentist's tools. With a wide range of shapes and edges, sculpting tools can be extremely convenient when it comes to filling gaps with putty or reshaping lost detail on a kit. Some of them, thanks to their pointed and sharp edges, can also, be used as scribers. Of course, they can be used for sculpting too—if you have the talent and skill!

MIXING PALETTES AND CONTAINERS

These inexpensive and often expendable tools are a great help when painting and finishing. You can find palettes in plastic, tin, wood, and even acrylic. They provide a clean and tidy way to mix custom paint colors. Containers come in many different sizes and are invaluable for storing excess custom colors, not indefinitely, but long enough for you to finish your project.

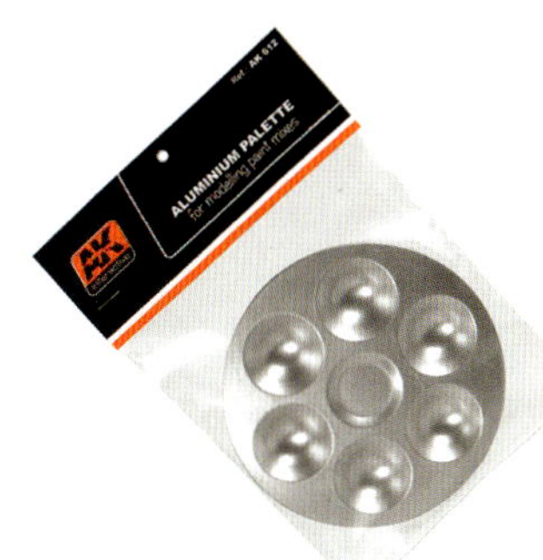

SPATULAS

Use spatulas to spread quantities of pastes and other semi-solid materials on a surface. They can be very helpful when building dioramas and filling large areas with plaster or other fillers. You should have a couple spatulas in different shapes to allow you flexibility in application.

In addition to those listed above, there is a vast variety of tools that can be useful to a modeler. They're not all purpose-made, and are often intended for crafts and hobbies other than scale modeling.

As your skills advance, you'll begin to recognize your need for a particular tool. It's a natural progression, and as you acquire these tools, you're ability to handle more demanding kits will improve too.

You'll find it helpful to keep empty water-bottle caps, plastic containers, and glass jars that you might otherwise recycle. Clean them out and store them for use as mixing cups, sorting containers, or dishes for soaking decals.

TIPS

- As your interest in scale modeling grows, you can invest some money and time on new tools.
- Most tools used in scale modeling can be found in hobby shops, hardware stores, and even art supply stores. Always check the quality of your tools.
- Don't acquire tools simply to own those tools; purchase tools that you'll use, not that you think you'll use.

Glues (types and uses)

Assembling a model kit dwells at the heart of scale modeling. To do this, we need glue.

Young and novice modelers, in a rush to build a model, will often to use any glue they have on hand, thus making assembly harder and the result unstable. In reality, you'll probably use a number of different glues, and you should know what they do.

PLASTIC CEMENT

Also known as liquid cement or liquid poly. This glue is the prime adhesive used in building plastic scale models. It reacts with and melts polystyrene plastic. The softened plastic parts actually merge upon contact, and, when cured, create a strong bond.

Plastic cement comes in different consistencies, from a thick gel to ultra-thin. The thinner the glue is, the quicker it hardens.

SUPER GLUE

Super glue (sometimes referred to as cyanoacrylate) was not initially made specifically for modeling, though some companies have developed purpose-made solutions. Super glues offer rapid-drying, ultra-strong bonds between nonporous surfaces. Some slow-drying versions can be found on the market, providing longer working time. Take extra care when using super glue because it can instantly bond skin to skin or skin to another material.

WHITE GLUE

Also known as acrylic or craft glue. Wood glue is a close cousin but has subtly different properties that make it ideal for working with wood.

White glue can be used on many different materials and offers strong bonds with porous materials such as paper and cardboard. White glue is nontoxic and can be cleaned up with soap and water if it hasn't dried. It has a rather long curing time, allowing adjustments and changes. It appears either white or clear when dry.

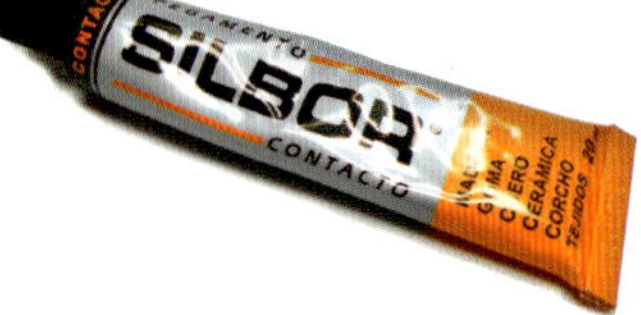

UNIVERSAL GLUE

Not the most convenient for constructing models, but it bonds diverse materials. This makes it a good choice for building dioramas and advanced modeling techniques such as flocking or adding hair and fabric onto large scale figures.

FIBERGLASS SCRATCH BRUSH

This isn't a glue, but it's a perfect tool to remove traces of any dried glue on your model.

Before gluing any parts together, make sure the parts are free of grease. You can clean the parts in a soap-and-water bath while still on the sprue. You can also clean parts with alcohol before assembling them. It's important that the pieces to be glued aren't forced and don't have tension between them. Be sure not to round or gouge edges; doing so reduces the contact area for the glue.

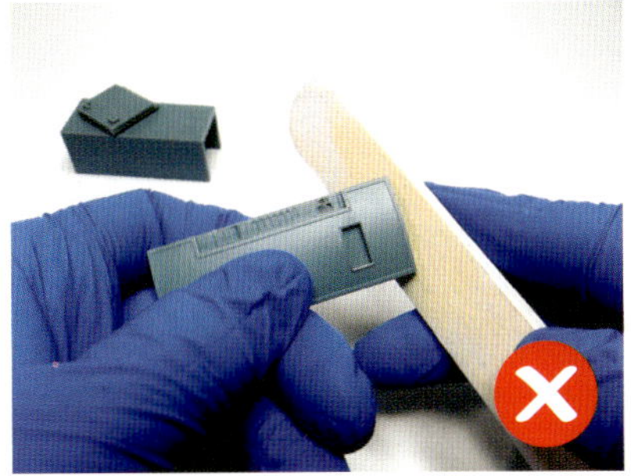

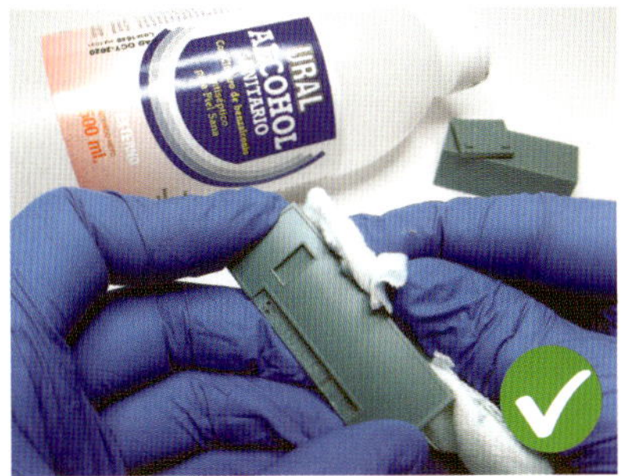

You should know the types of joints you can make when gluing two pieces. Knowing these basic joints can be useful when going over assembly instructions and may make a seemingly confusing diagram clear.

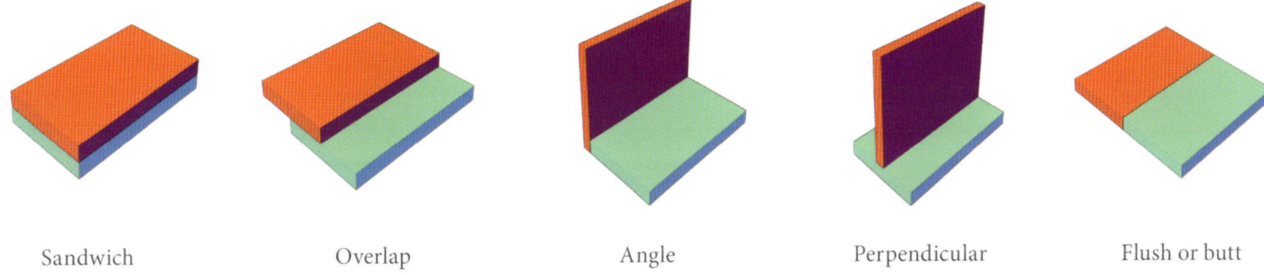

CAPILLARY GLUING

Capillary gluing is probably the most popular method for gluing along seams. Capillary gluing takes advantage of the tendency of thin liquid cement to flow along the space between two parts placed next to each other. As always, the closer the fit, the stronger the joint.

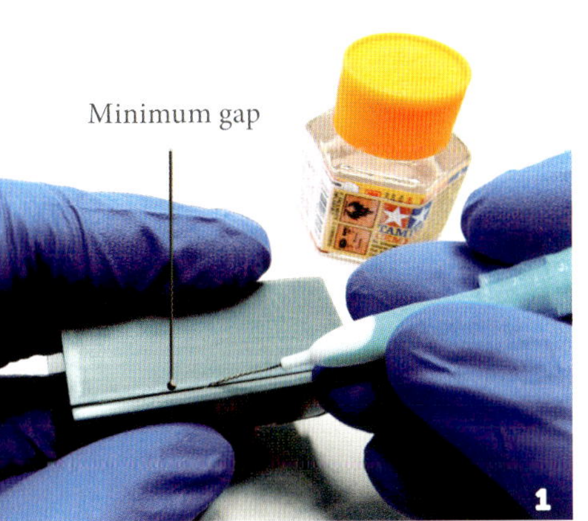

1. The two surfaces to be cemented need to have enough room to introduce the applicator and allow the adhesive to flow along the joint.

2. After about 30 seconds we close the joint, applying enough pressure to close the gap. Be careful and control the amount of glue you flow along the seam. Too much and you can soften adjacent areas you didn't mean to; not enough and the joint will be weak.

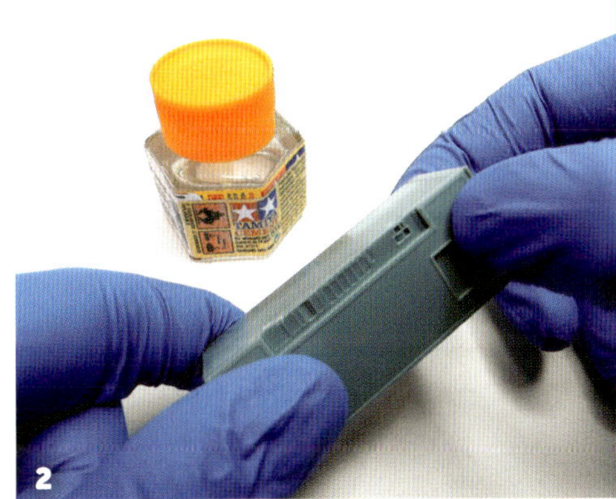

TIPS

- Use clamps when applying a slow-drying glue. They will hold the pieces firmly together until the glue has dried.
- Some glues are toxic and can cause injuries if they come into contact with skin or eyes. Use gloves and eye protection and have eye wash on hand in case of an accident.
- A brush, photo-etched-metal applicator, or toothpick can help make your glue applications more precise.

Putties

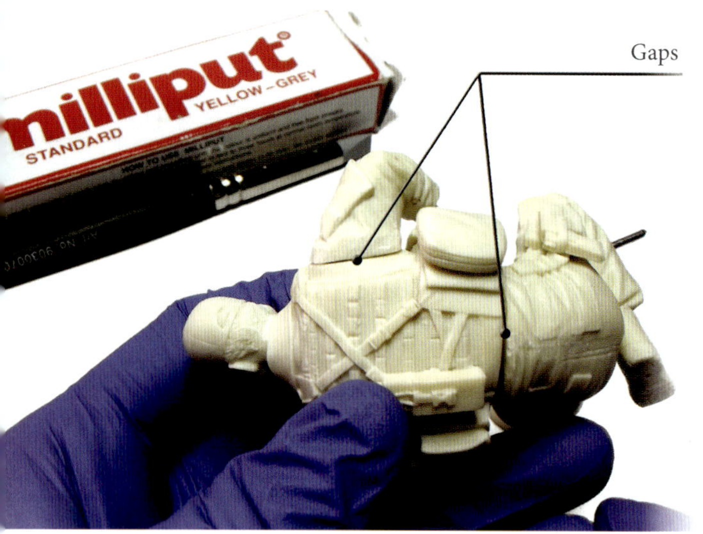

After completing an assembly session, it's always advisable to apply a coat of primer to your model. The primer not only unifies the model's surface and prepares it for paint, but provides a good—and sometimes last—chance to check the quality of your work. It reveals gaps and misalignments, and fit inaccuracies of the kit itself or of modifications you may have made.

Whatever the cause, such defects must always be corrected. Often, putty is the first material modelers reach for to make such fixes.

There are many different types of putties to choose from, some specially made for scale modeling, others for general purposes. Putty is an air-curing material that hardens to a consistency that can be sanded or carved—perfect for filling gaps. Some putties can be shaped up to considerable volume, allowing you to form complex shapes.

In general, modeling putty can be diluted with acetone. Acrylic putty is becoming more popular, and it can be thinned and worked with water. Putty tends to shrink as it cures. The less a putty shrinks, the better it fills the gaps with a single application. If a putty shrinks considerably, you'll have to repeat the filling process as many times as needed to evenly cover the area.

Applying a ready-to-use putty is generally straightforward: You place a quantity of putty on a palette or other smooth, nonporous surface (1).

1

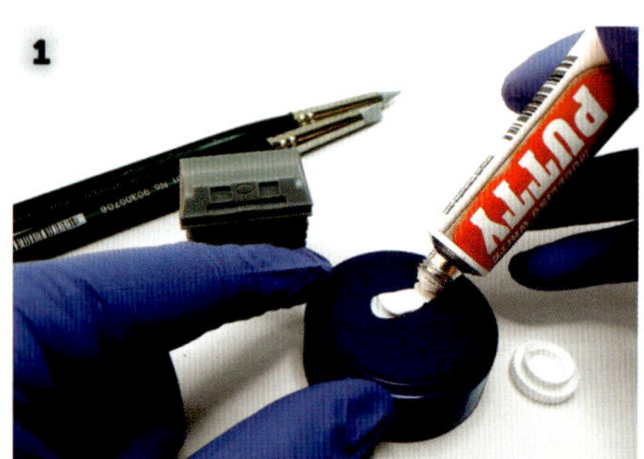

2

Spread it over the areas needing correction with a spatula, brush, dental tool, or even a toothpick (2). Wait the curing time specified by the manufacturer to make sure the putty has dried completely.

When ready, you can sand the putty with sandpaper of gradually finer grits until you achieve a smooth and even surface (3). While sanding, be sure not to remove any details you wish to keep on the model.

3

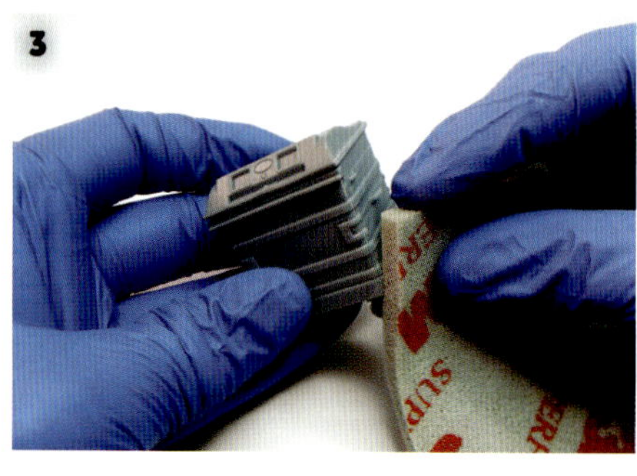

TWO-PART EPOXY PUTTY

As the name implies, two-part epoxy putties come with an epoxy resin putty and a hardener putty. Epoxy putty can be used to fill wide gaps and can be shaped into forms.

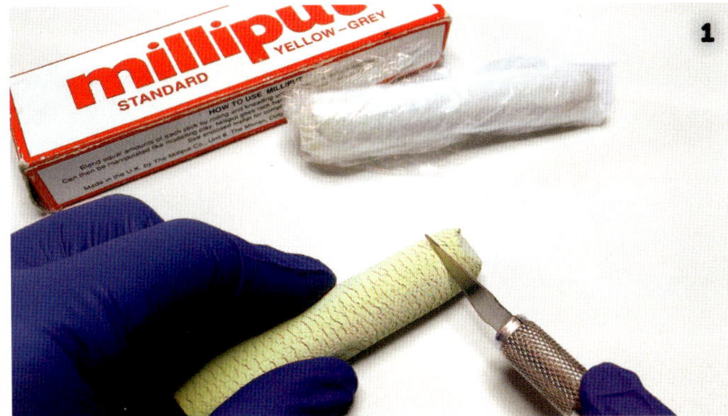

1. Typically, you cut and mix equal parts of the two components, but always check the manufacturer's instructions.

2. Knead the two components together until they make a uniform mixture. Often, the parts are different colors that become a consistent single color when completely mixed.

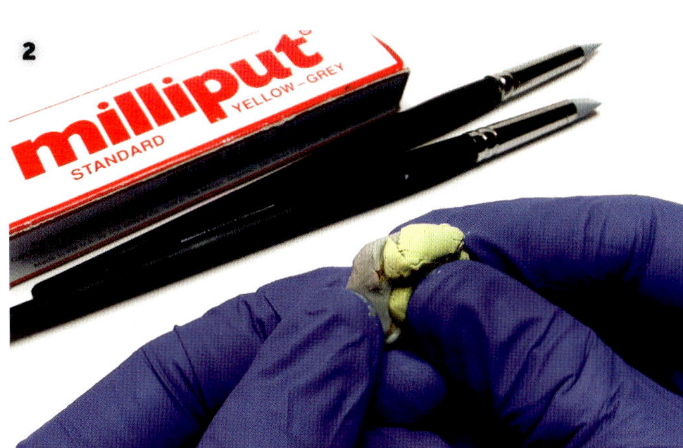

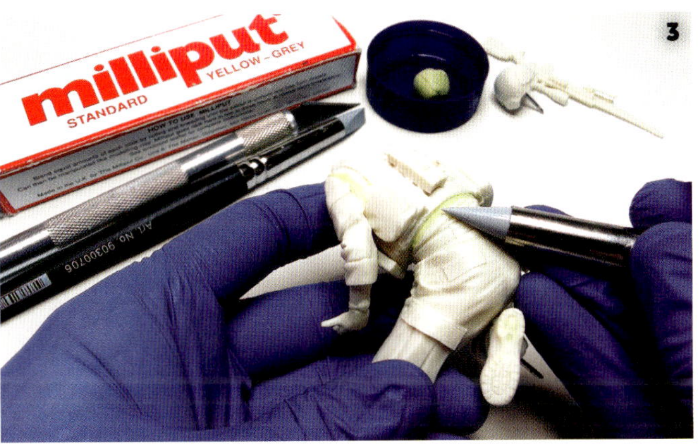

3. Fill the gaps and imperfections on the model, taking care not to cover nearby details. Epoxies usually harden ultra-strong, making it impossible to remove from fragile details. Use a toothpick or a brush, (either rubber or conventional), to apply the putty. Always have some solvent or water (as indicated by the maker) on hand to remove or shape the putty.

4. If you're using a putty with a stiff enough consistency and an appropriate curing time, you can use it to add or complete missing parts, shapes, or details on your model, such as ribbons, fabrics, shields, and rivets.

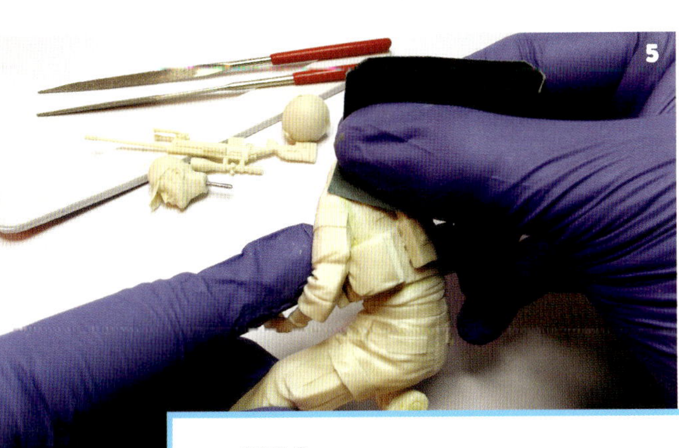

5. Finally, let the product dry for the time indicated by the manufacturer. After the epoxy has thoroughly set, it can be filed, carved, and sanded as needed.

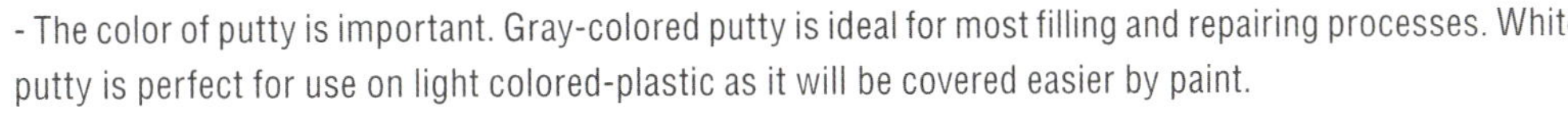

TIPS

- The color of putty is important. Gray-colored putty is ideal for most filling and repairing processes. White putty is perfect for use on light colored-plastic as it will be covered easier by paint.

- When creating small details with epoxy putty, it's a good idea to work on a nonstick surface, such as wax or parchment paper, to keep the putty from sticking to your workbench.

Sheet styrene, strips, and more

While building a model kit, you'll sometimes find parts or details that you would like to alter, improve, or even add, in order to improve the model's final appearance. In other cases, you might like to build scenery to more effectively display your model. For all those different cases, sheet, strip, rod, and tube styrene are the materials for you!

The variety of styrene materials on the market is staggering. You can find sheets in thicknesses from .005" to .125" or more. Shapes range from I-beams to Z-channel, tubes and rods, and strips of different widths and thickness. You can even find premade styrene siding, brick facades, and roadways.

With styrene components, a little planning, and your growing modeling skills, you can build almost anything. This is called scratchbuilding—building something from scratch. Now, let's see how to work correctly and effectively with plastic sheets, strips, and shapes.

1. The first thing you need to have when making parts from scratch is a plan. Always take into account the scale you're working in so everything remains proportional to your model or diorama. Correctly size your parts by making a sketch. When working on projects such as updating or backdating an airplane or tank, or when building a specific type of vehicle or item, closely follow your plans or references to avoid mistakes and wasted material.

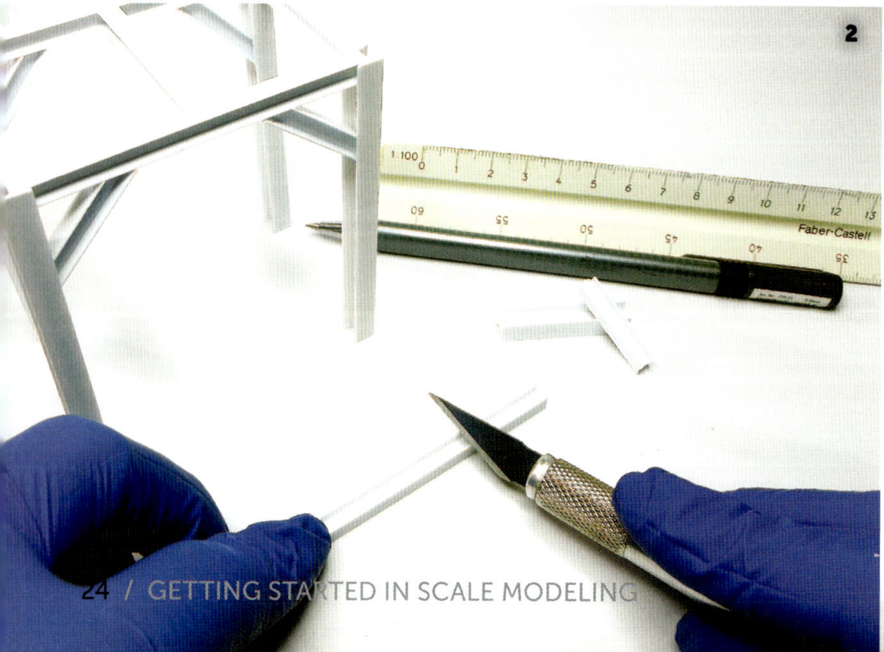

2. Once you're certain what you're going to build, and its exact shape and size, you can proceed to cutting the parts.

Styrene, being a fairly soft material, can be cut with a hobby knife or, when necessary, a razor saw. Use a metal ruler as a guide to make straight, clean cuts.

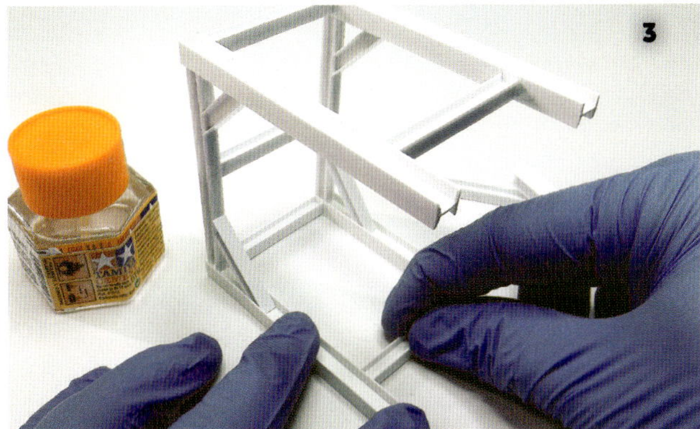

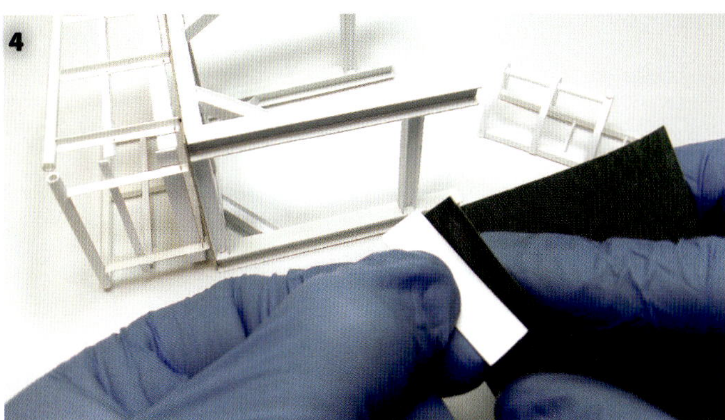

3. To assemble styrene, use plastic cement, super glue, or epoxy if the joints need extra strength and durability. White glue won't work for this type of construction. When applying any type of adhesive to styrene, remember to be careful and apply it only on the areas to be joined; it can damage the plastic and create undesired textures.

4. Once your pieces have been glued together, give the model a gentle sanding with very fine sandpaper. This will remove any leftover glue and smooth irregularities.

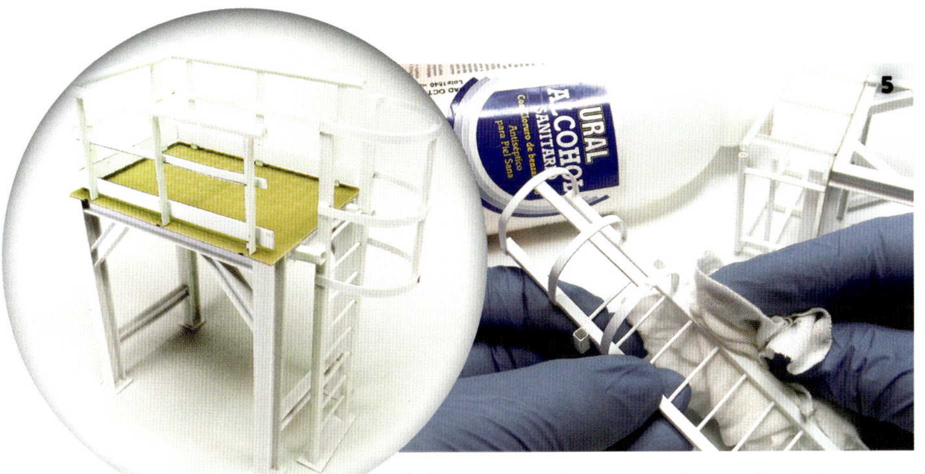

5. Clean the assemblies with alcohol to remove grease, dirt, and marker or pencil marks left from the design process. Make a final check of all the glue joints. Reinforce or fill any joints that require it.

6. Styrene can be engraved or scribed. For example, you can create realistic wood textures with a wire brush or a fine scriber. Treating a styrene surface with liquid glue, then gently tapping a hard brush onto it, can provide a nice cast steel texture.

TIPS

- Make a thorough study of the information and reference of the item you want to build in order to stay as close as possible to the real thing or what you envision.
- Super glue provides instant and strong adhesion, but use it wisely when working with styrene. The bonds can become brittle over time.

Decals

Before beginning work on a project, you'll probably already have an idea of how you want the model to look. Although most of the painting will be done by you, there are certain symbols or markings that are too complicated to re-create by hand, regardless of your skill. Logos, military and civilian symbols, advertisements, badging, and more, have patterns and shapes that need to be precisely depicted on our model for it to be a replica of the real thing. Making things harder, often markings must be repeated multiple times on a model.

To solve this problem, all kit manufacturers offer decals. Decals are scale copies of the actual markings, printed on film or special paper that allows us to transfer them onto a model. In general, model kits include a decal sheet to complete certain versions of the subject. But there are many aftermarket decal sheets available to allow you to finish your model as a version that may not be provided in a kit. You can also print your own decals with your printer. However, the quality will likely be lower than professionally printed decals.

DRY TRANSFER DECALS

Dry transfer decals can be applied without water or solvents. They're printed on special paper or plastic backing. You place the transfer over the desired location on your model with the backing side up. You transfer the decal by burnishing the backing with a sturdy implement. The contact side of the decal has pressure-sensitive adhesive making it cling to the model's surface.

Before you apply the decals, prepare the surface with a coat of gloss varnish. The gloss will allow the decal to adhere better.

Cut each decal you want to transfer from the decal sheet.

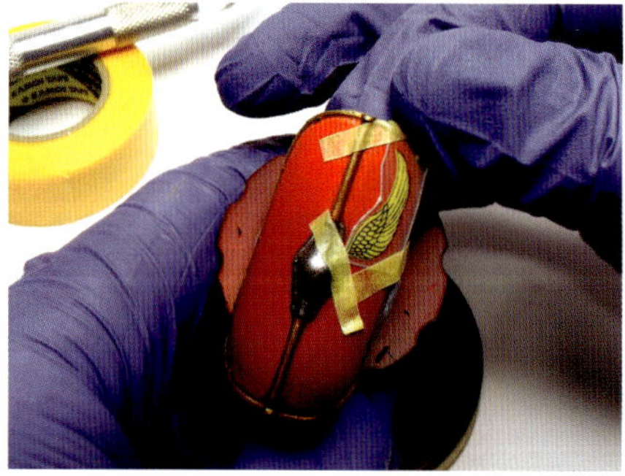

Position the decal and attach it in place with masking tape—you don't want it to move.

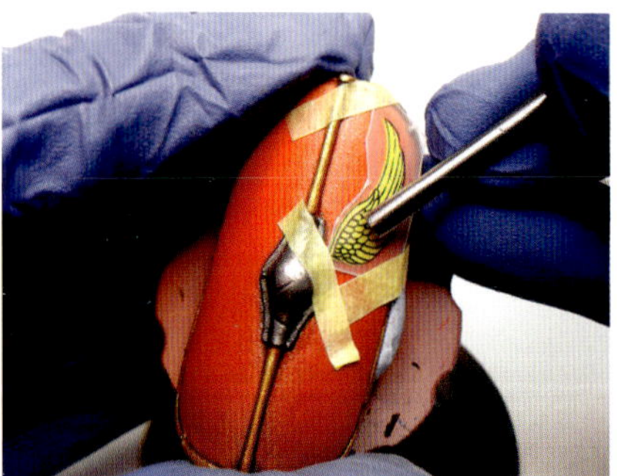

Gently but firmly burnish the decal backing with a rigid tool, making sure that all of the decal transfers from the carrier sheet to the model.

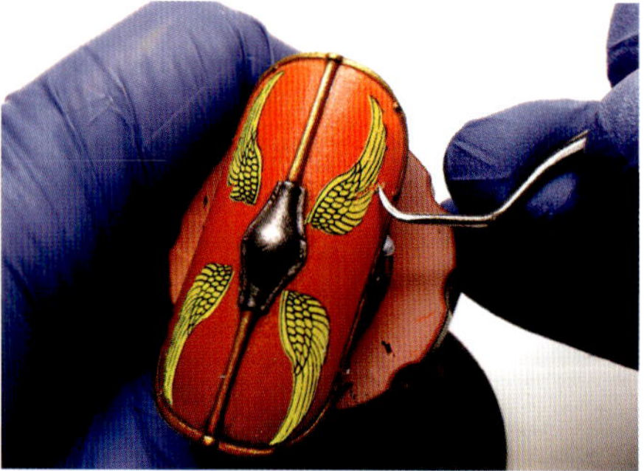

Once the decal has been transferred to the model's surface, you can create minor scratches or other traces of wear to increase realism.

Minor details can be painted directly onto the decal to customize or correct it. Depending on the subject, finish the operation with a coat of flat or gloss varnish.

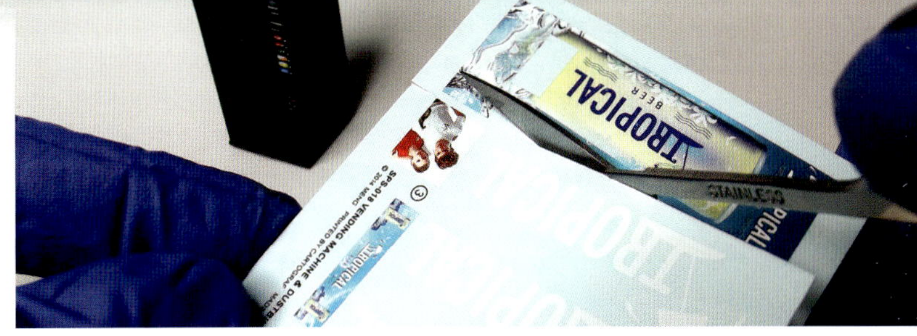

WATER SLIDE DECALS

Water slide decals are printed on a carrier film attached to paper backing. The film is released from the backing by moistening it with water. Once placed in position, they stick to the model's surface by means of an adhesive layer. Decal softeners and setters can enhance the quality and effectiveness of water slide decals.

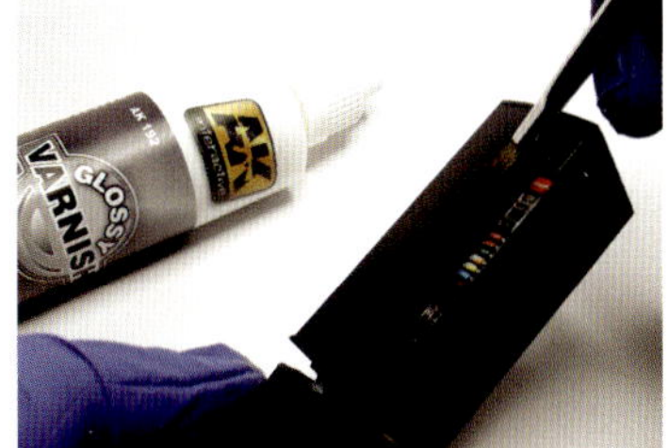

GLOSS VARNISH

The surface of flat paint is very rough, making it hard for water slide decals to adhere well, trapping air under the decal. This causes the decal to have a distorted or silvery appearance. To avoid silvering, always apply a coat of gloss varnish on the area where you'll apply your decals. Sometimes, it's easier to coat the entire model using an airbrush than spot coat specific areas by hand.

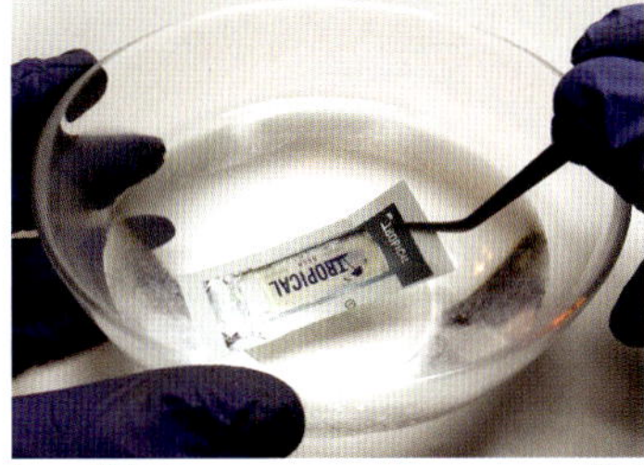

WARM WATER

Cut out the selected decal from the sheet and dip it in a shallow dish of warm water for a few seconds. Don't let it soak or you run the risk of the decal floating off the backing paper in the water. Set the decal aside for 20 or 30 seconds to allow the adhesive to activate, then hold it over the marking's location with tweezers. Gently push the decal in place with a toothpick or similar tool while still holding the paper with the tweezers. Using a cotton swab, gently roll away the excess water and simultaneously press the decal onto the surface.

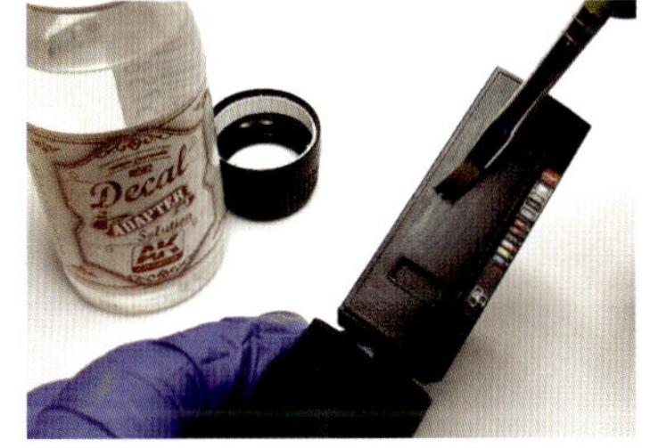

DECAL SETTERS AND SOFTENERS

Sometimes a model's shape makes it difficult for a decal to adhere properly. In those cases, you should brush the model's surface with setting solution. Place the decal over the setting solution as described above.

Once you're pleased with the decal's placement, brush more of the setting solution over it and gently press it with a cotton swab. Once the setting solution has dried, brush a coat of decal softener over the decal to get a painted-on look.

Once completely dry, water slide decals can be treated with various weathering and fading techniques. Scratches, shadows, and even missing details can easily be applied with a fine brush or hobby knife. Seal decals with flat, satin, or gloss varnish as appropriate to the model.

TIPS

- If air bubbles appear beneath our decals, poke them gently with the tip of a sharp hobby knife or sewing needle, then apply setting solution.
- Weather after placing the decals in order to blend them with the overall look of your model.

How to use paint strippers

An issue a modeler will surely come up against will be the need to remove paint from a model. This happens for a variety of reasons: unhappy with how the project ended up; mistakes during painting that require starting over; improved skills. Painting over a previous paint job can be tempting, but it's not a good idea. Successive layers of paint build up, adding thickness and obscuring details and features such as panel lines, rivets, and facial characteristics.

There are many options for stripping older paint jobs depending on the type of paint used (enamels, acrylics, lacquers, etc.). Turpentine, acetone, brake fluid, and alcohol are just a few. You can employ them all similarly with acceptable results. However, in most cases, they will not completely remove all the paint, particularly from engraved, etched, or textured surfaces. Below, you can see how to use denatured alcohol to remove acrylic paint from a metal figure.

1. Pour enough alcohol to cover the model in a glass container.

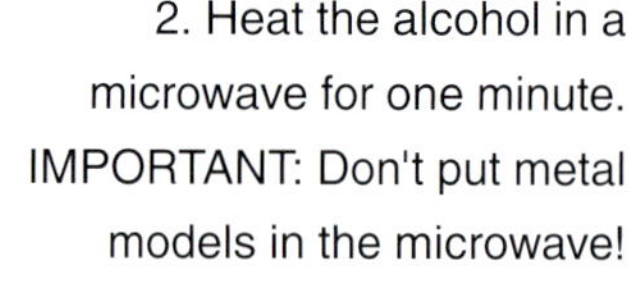

2. Heat the alcohol in a microwave for one minute. IMPORTANT: Don't put metal models in the microwave!

3. Once the alcohol is heated, place the figure into the liquid and leave it submerged for five minutes. Whenever you use heated liquids, it's always a good idea to wear protective gloves and safety glasses to minimize the chances of burns or other injuries.

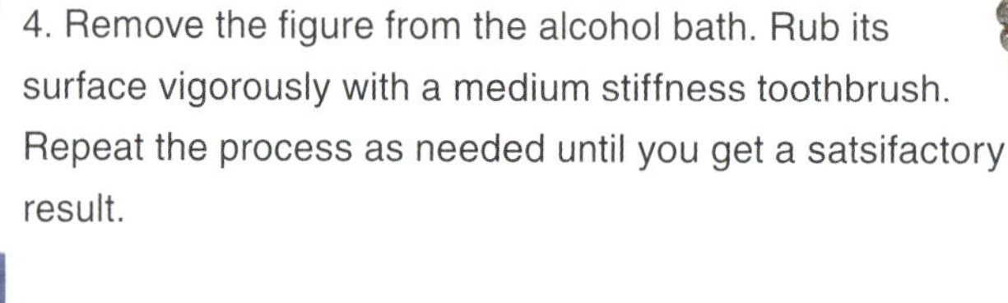

4. Remove the figure from the alcohol bath. Rub its surface vigorously with a medium stiffness toothbrush. Repeat the process as needed until you get a satsifactory result.

Even after such a drastic operation, some paint may still be visible in deep cuts and fine details. You can try to remove them with a sharp tool, but this can damage the figure. At this point, clean it with water and the toothbrush and call it good.

You can also use stripping products made with modelers in mind.

1. We will need a container big enough to fit our model or parts.

2. Pour enough paint stripper into the container and place the model inside. If the model or parts are very large, you can apply a generous wash of stripper on the surface using an old brush, simultaneously rubbing the surface, focusing on deep recesses and details.

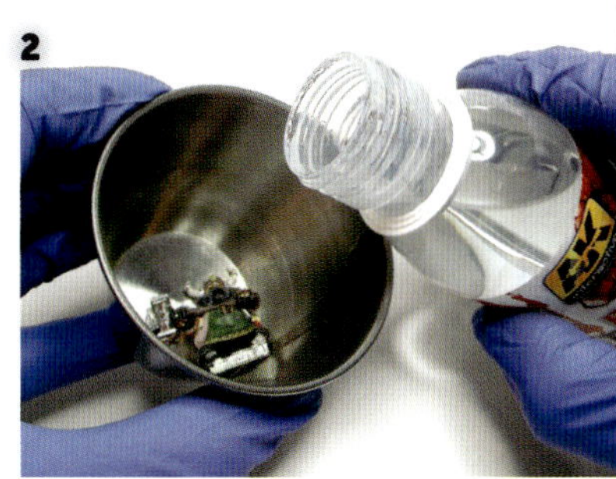

3. Here, the model is submerged in the paint stripper.

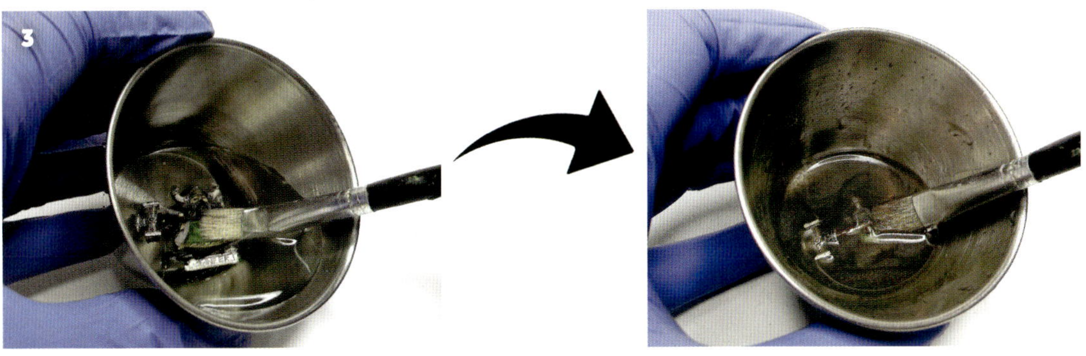

4. After some time has passed, gently rub the surface of the model using a stiff-bristled toothbrush. Focus on recess to remove any unwanted paint. This is quite a dirty job, so be sure to cover your workbench.

With only a few minutes' work, you can effectively strip old paint from your model and have it ready for a new finish. If you're not happy with how that one turns out, you can always repeat the process.

TIPS

- All methods mentioned above are done with very strong products that can cause damage to work surfaces, tools, or other models if abused.
- Whenever working with solvents, have clean water nearby and use gloves and eye protection.

In almost all cases, when we get a scale model kit, we will find its pieces disassembled, spread across sprues, ready to be built. Many modelers consider the construction process a boring, unavoidable stage that has to be passed in order to get to the more creative and fun part: painting! Yet, model kits are meant to be built, aren't they?

So, we always have to keep in mind that the assembly process of a project is also part of the fun of our hobby and we can enjoy it too!

The pieces comprising a model must be treated correctly in order for the result to be satisfying. In this chapter, you'll learn the basics necessary to build a scale model, from using tools, to optimizing your building process. Furthermore, you'll learn some techniques for improving kit details.

Pay close attention and try to be as thorough as you can! As your skills evolve, you'll find your models looking increasingly more realistic.

2. Basic building.

How to remove parts from the sprue

Most scale models, no matter whether figures, vehicles, or buildings, usually come disassembled in multiple pieces. For plastic kits, the pieces are fixed in frames called sprues. In resin kits, parts come attached on a block of material called a runner. In both cases, the first thing to do is to remove the pieces necessary for the current assembly stage and only that stage—you don't want to lose parts.

For now, follow the instructions and use them as a guide for the order of assembly. With experience you can begin to deviate.

Pieces on the sprue

Pieces with runners

Cutting the parts loose always leaves behind tiny portions of the sprue. These small sprue attachment points, or burrs, will make the surface irregular, spoiling part fit during assembly. You can make quick work of burrs, but you need to be meticulous and careful.

The main tools you'll need for removing parts from the sprue and then cleaning them up is a pair of side-cutting pliers and a sharp hobby knife, preferably with a No. 11 blade. Remember, don't snip the sprue attachment point flush with the part. This can damage the part or create distortion you'll have to repair later.

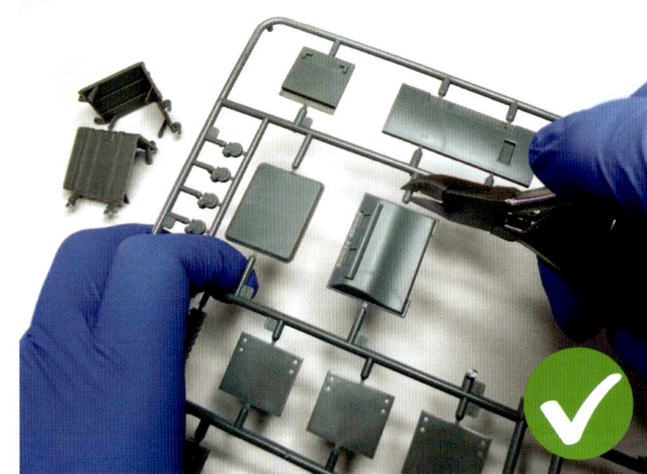

1. Remove the burr with a hobby knife. Try to be very precise and remember that the knives cut best when you use a slicing motion, not just pressure.

2. Go over the whole area with fine sandpaper or a sanding stick. If the part is metal, use a file to remove excess material.

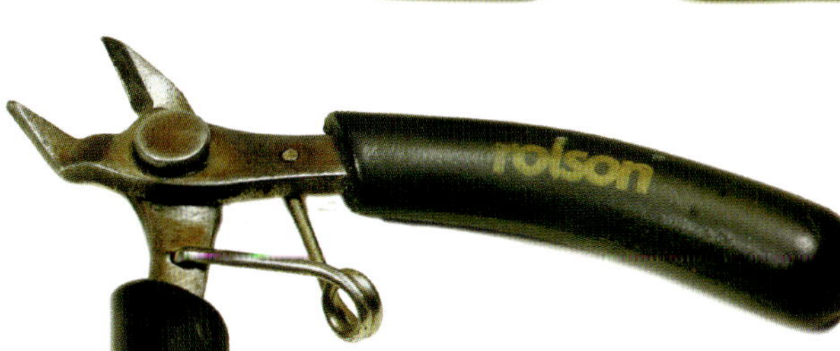

Depending on the type and quality of the plastic, you may see some discoloration after sanding, especially on prepainted models. Don't worry. It will get fixed.

TIPS

- Always use good quality tools to obtain the best results. It's preferable to have a few good tools rather than many low-quality tools.

- Thoroughly check your work and be patient.

How to remove mold lines

In almost every kit, regardless of material or subject, you will notice the parts provided have minor flaws from the production process.

There are two types of production marks: mold lines and ejector-pin marks. Both need to be corrected before building your model.

MOLD LINES

Mold lines appear at the point where the parts of a production mold meet. A mold line appears as a thin line that runs along the part. If you're lucky, it's on an edge that's easily hidden, but that's not always possible.

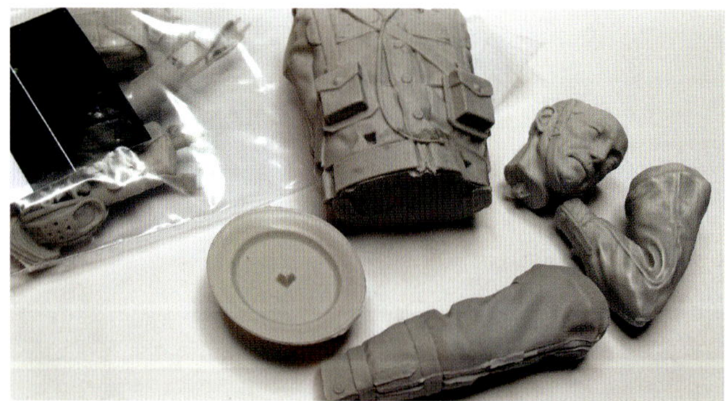

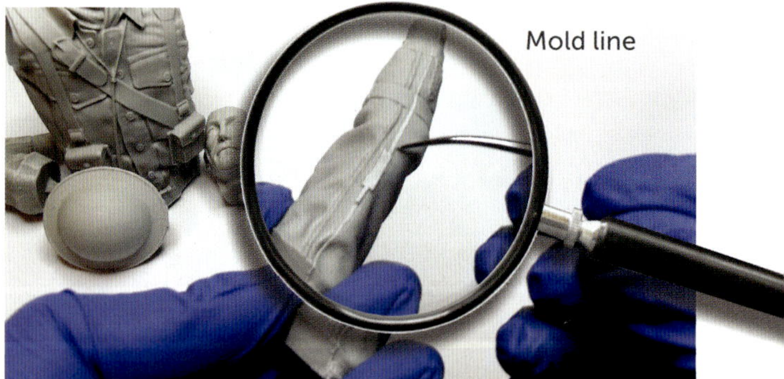

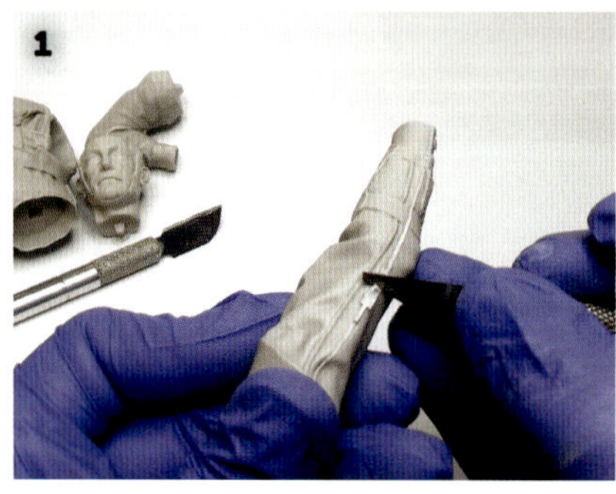

1. Use a hobby knife to shave away the excess plastic. If the mold line is particularly fine, you can use the back of the blade to remove the mold line. Place the blade perpendicular to the mold line and gently scrape.

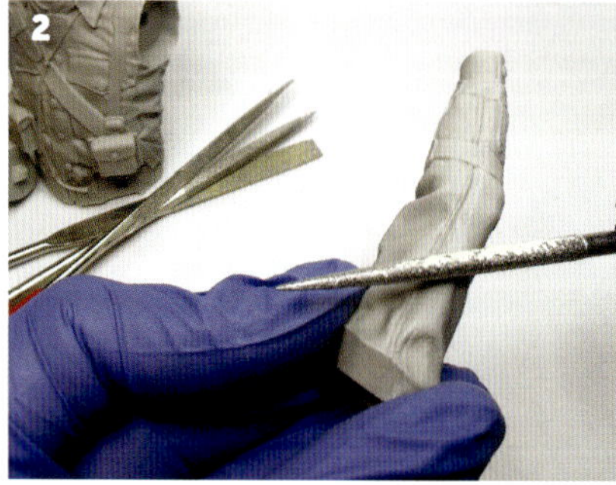

2. Depending on the shape of a surface, a round or flat file will remove any remaining mold lines and blend your work into the surrounding contours.

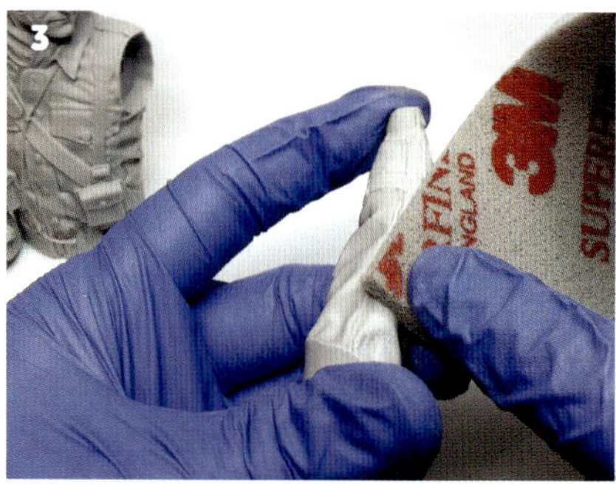

3. Go over the area with progressively finer-grit sandpaper to smooth the area completely before cleaning it with a corner of a clean cloth and a small amount of denatured alcohol.

In some kits, mold lines may also appear in the form of a level mismatch. This is caused by old or misaligned molds. To correct this flaw, you'll need to use putty and shape the area with sandpaper.

EJECTOR-PIN MARKS

Ejector-pin marks are made when the parts are ejected from the mold. Luckily, they are usually found on the underside of parts or in places where they'll be hidden from view. However, that's not always the case. When you know an ejector-pin mark will be seen on the final model, it needs to be repaired. Ejector-pin marks don't appear on resin or metal parts due to a different manufacturing process.

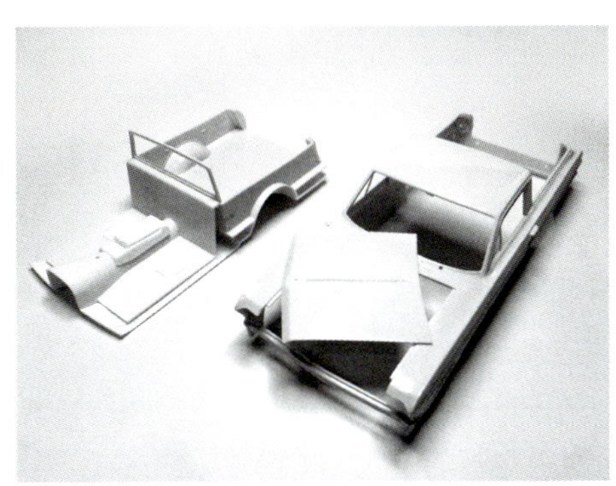

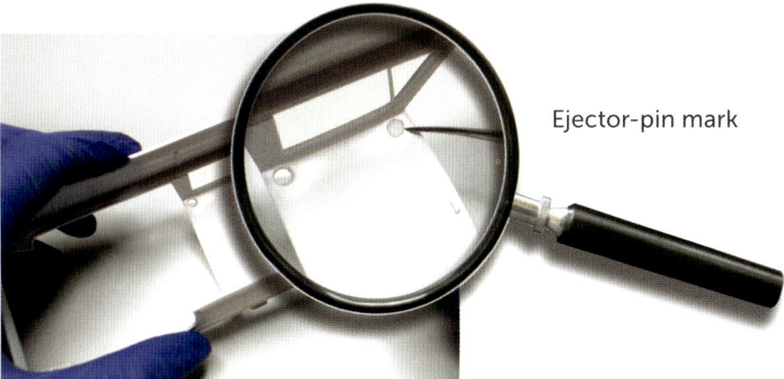

1. Some ejector-pin marks can have an excess of material. Carefully remove the material with a hobby knife.

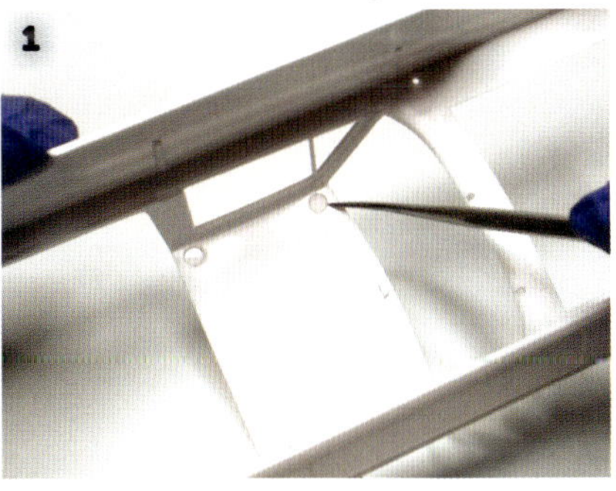

2. Smooth the surface with a file or sanding stick.

3. Fill depressions with putty. Remember, putty shrinks as it dries, so add a little extra.

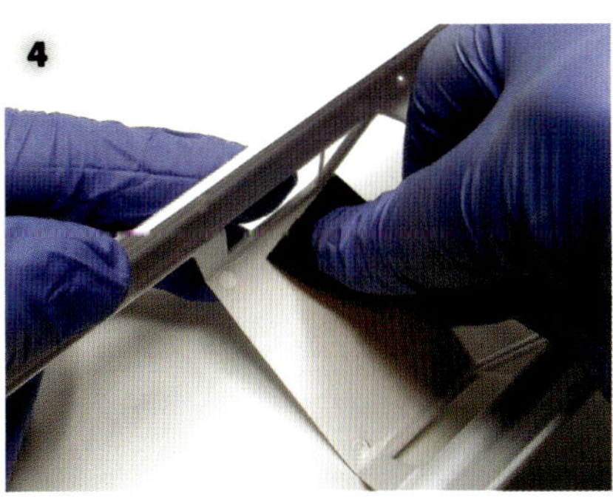

4. Once dry, use sandpaper to remove excess putty and move to progressively finer sandpaper until you obtain an even and smooth surface.

Both mold lines and the ejector-pin marks, if not treated properly, can mar the final appearance of your model. Locate and treat them during the building stages, because it will be almost impossible to do so after it has been put together or after you've painted.

TIP

- Locate all mold lines and ejector-pin marks during assembly. Each one has to be corrected unless you're sure it will be hidden on your model. If you're not sure, better to be safe and correct the line or mark.

Clear parts

Clear parts simulate the transparent areas of a vehicle or building, like windows, windshields, and headlights. Usually, they're made of injected transparent plastic, although you will occasionally find sheets of transparent styrene or acetate in a kit.

Transparent parts must be carefully treated in order to remain clear and maintain their luster. First, plastic cement and its vapor can damage clear parts. Enamel colors or aggressive solvents can make them lose their transparency, too. This undesired effect is called fogging or crazing. If damaged, destroyed, or lost, clear parts are difficult to reproduce. Vacuum-forming is a solution, but it's extremely complicated and difficult for beginners or untrained modelers, and the resulting part will probably not be as good as the original.

If you scratch a clear part, all is not lost, but the process is beyond the scope of this book. For your purposes, be mindful of your clear parts and keep them protected in a resealable plastic bag. Leave them alone until you're ready to finish them and attach them to your model.

The clear part on this windshield has been deliberately scratched with the tip of a hobby knife blade to simulate the impact of a bullet. A micro drill bit made the small hole in the center to complete the effect. Pigments provide the dust. Pigments offer a simple and safe way to depict dirty windows because of the control they allow in their placement; if you don't like the effect, the pigments are easily removed.

Precise masking is important when you paint clear parts, such as airplane canopies. You can find many different kinds of masking tape on the market, as well as masking paper for larger projects, and liquid mask that you can brush on.

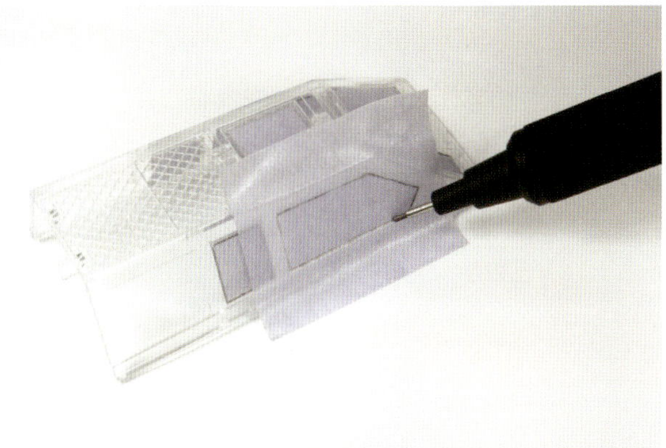

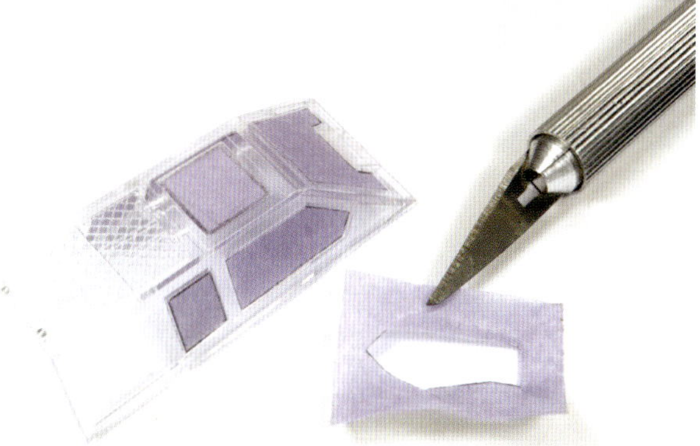

With masking tape, cut a small strip and apply it to the area to be masked. Run a mechanical pencil along the edge and then use a sharp hobby knife to make the cut.

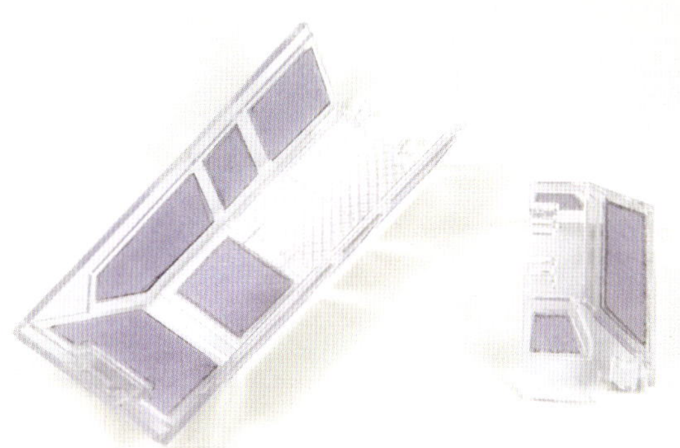

Some clear parts like headlights, taillights, indicators, anti-glare glasses, or periscopes may need to be a particular color, but remain transparent. This can be achieved using clear or transparent paint. These colors can be applied with a paintbrush or an airbrush, as you can see in the example, below.

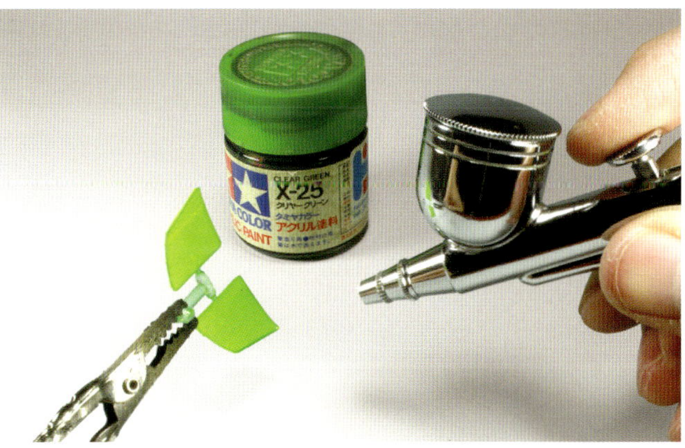

TIPS

- Do not keep masking tape, liquid mask, or other masking material on parts for prolonged periods of time. The adhesive may damage or fog your parts.

- It's better to mask a clear part entirely even if you intend to paint only a small part of it. Overspray can destroy the rest of your part.

- Pledge Floor Gloss floor polish can increase the transparency and luster of a clear part. It also protects the surface from scratches.

Photo-etched parts

The physical limitations of plastic and resin make molding certain small, thin details impractical or impossible. The answer for many of these tiny details has become photo-etched-metal parts. Originally found as aftermarket upgrades, many kits now come with photo-etched parts included.

Usually made of brass, photo-etched parts are made via a chemical process and create precise parts that come attached to sheets called frets. Photo-etched parts can provide a high level of detail, enhancing the final appearance of your model.

Normally, photo-etched parts require priming before painting because their smooth metal surfaces offer little for paint to grip. There are more and more prepainted photo-etched parts available, however, allowing for quicker assembly, but still leaving you with plenty of options for weathering.

1. Photo-etched parts can be extremely fragile, so be careful with them. You can use the back of a hobby knife blade or set of sturdy scissors to remove the parts from their fret. Try to cut the retaining tab as close to the part as possible to minimize cleanup. Diamond files work very well on photo-etched metal. Sometimes you'll find a fret has plastic film attached to it. Simply peel it off before cutting snipping the parts free.

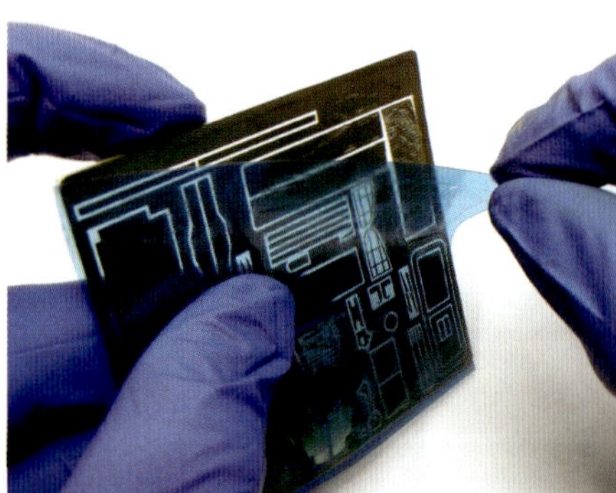

2. Before applying glue to a photo-etched part, rough-up the contact area with a file to give the adhesive some texture to grab. If the part is large enough, you can use a pair of tweezers to handle it during filing. For smaller parts, you should consider filing the contact surfaces while the piece is still attached to the fret in order to avoid unwanted folds and breaks.

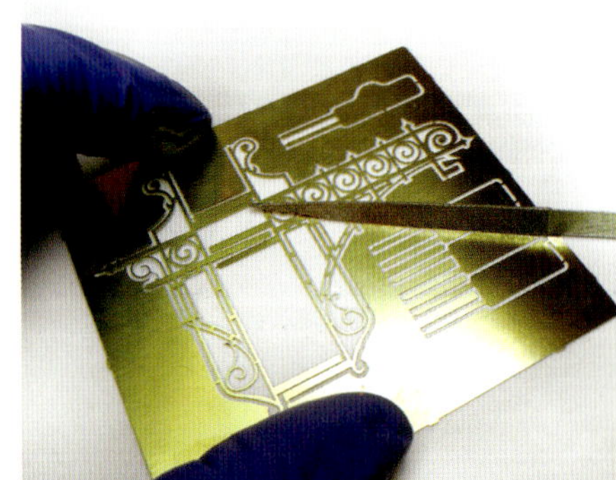

3. Photo-etched parts often require shaping before they're attached to the model. Most photo-etched parts come with fine grooves as a guide showing you the exact spot where it should be bent. Special tools for photo-etched parts have been developed, making it even easier and safer to correctly angle pieces.

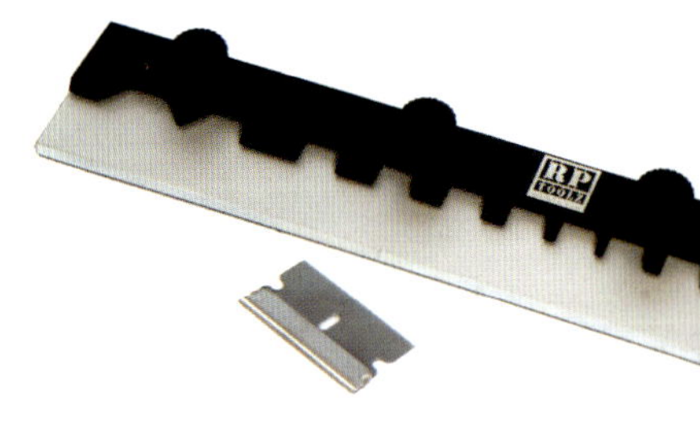

4. If you don't have a bending tool, you can do the work with a pair of flat pliers. Just place folding line of the piece on the edge of the pliers, grabbing the rest of the part with a second set. Bend the piece carefully until the correct angle is obtained. Always check that the bend is even along its full length. Once your model is fully assembled with all the photo-etched-metal parts in place, you must apply a layer of primer to help the paint to adhere to the metallic surface.

Modern photo-etched parts have been a substantial step forward in scale models, enhancing overall kit detail and quality. Without them, some parts would be almost impossible to accurately depict. Prepainted photo-etched parts further improve these details.

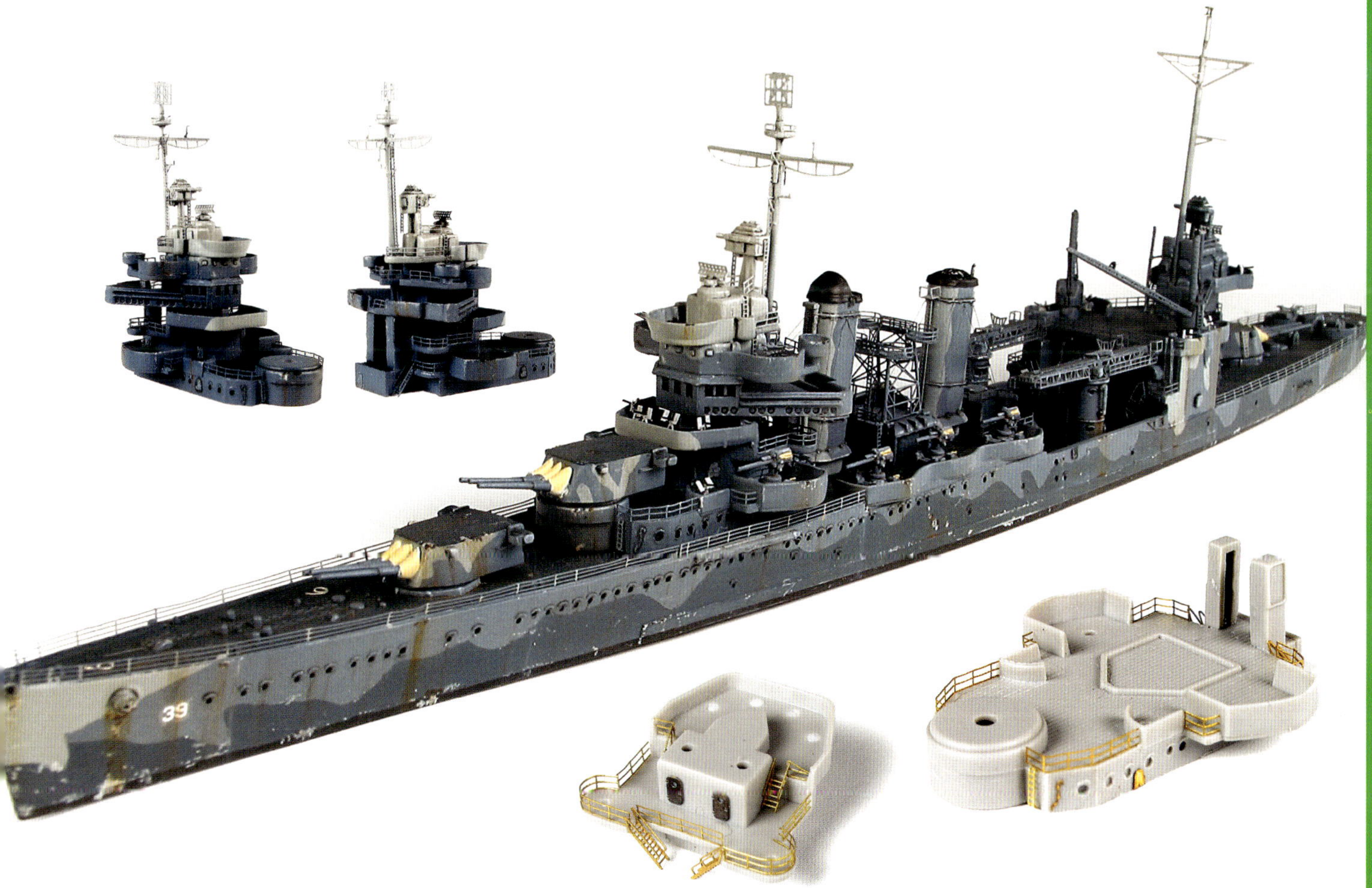

TIPS

-Clean super glue from metal glue applicators with the flame of a small candle.

-For small, fragile photo-etched parts like ship railing, glue one end of the part to the model and slowly work your way around the model, making sure to shape the part as you go.

Scribing

Sometimes, particularly with older kits, you'll find the panel lines are raised rather than engraved. Other times panel lines will be missing altogether. To improve the accuracy of such kits, you can scribe panel lines and similar details.

Scribing requires a sharp tool, but can be accomplished with a hobby knife in a pinch. The better option is to employ a scribing tool (also called a scriber). Drag the tool along the plastic or resin surface following a ruler, purpose-built templates, or self-adhesive labelling tape available from office supply stores. Lightly drag the scriber along the edge of your scribing guide. Repeat the process to deepen the line for the scale effect you're looking for. For example, scribing a panel line on a 1/72 scale kit must be hair-thin to be realistic, so you don't want to make it too wide. A similar panel line on a 1/32 scale kit requires a wider, deeper groove for the same effect. You also may have to re-scribe panel lines lost to extensive sanding.

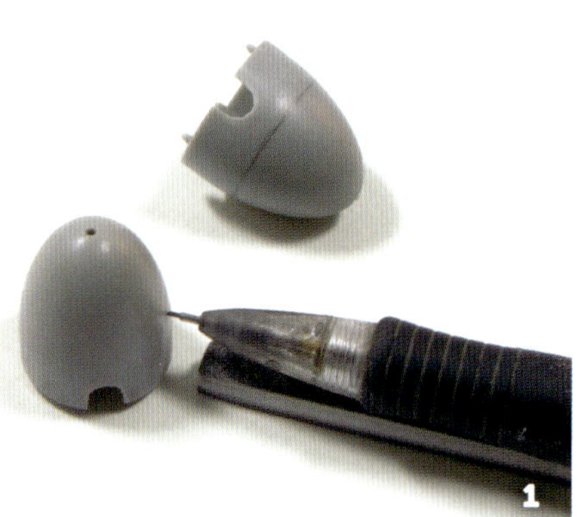

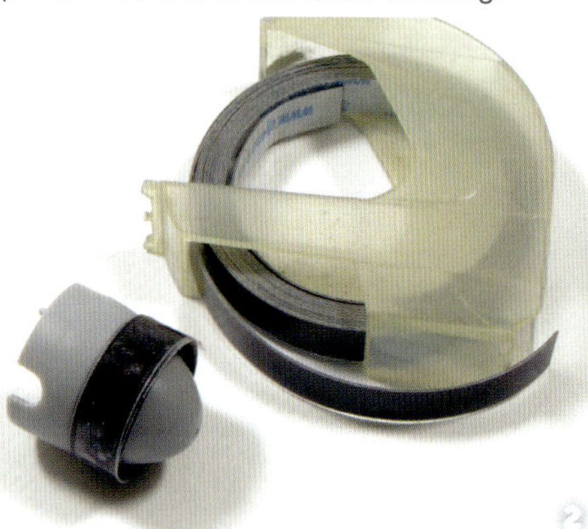

Scribing a straight line or a specific shape on a surface can be tricky on a part with curves or complex surfaces.

Start by drawing a pencil line on the surface to mark where the panel line will be (1). In the example, you can see how label tape can be used to create a guide on a propeller nose cone, or spinner (2). Our scribing tool must be sharp to engrave the plastic. The first line can be the hardest to align. Mistakes happen. Go slow and get the line as perfect as you can. Repeating the process following the initial groove should prove easier (3) If you do make a mistake, fill the unintended line with either fine putty or super glue. Let the filler set and then sand it smooth.

After you've finished scribing, lightly sand and polish the part to blend the engraved detail. Some experienced modelers flow liquid plastic cement into the scribed detail to soften the effect.

You may find parts with damaged or missing raised rivets. You can find replacement rivets in different scales in resin, photo-etched metal, and even as adhesive decals like what's shown here.

A specialized rivet maker can be used to make recessed or flush rivets. Rivet tools come with interchangeable heads with evenly spaced teeth in an assortment of diameters and tooth sizes. To be accurate and effective, rivet makers require a ruler or template as a guide. New rivets can be smoothed with liquid plastic cement to blend them with the model's surface.

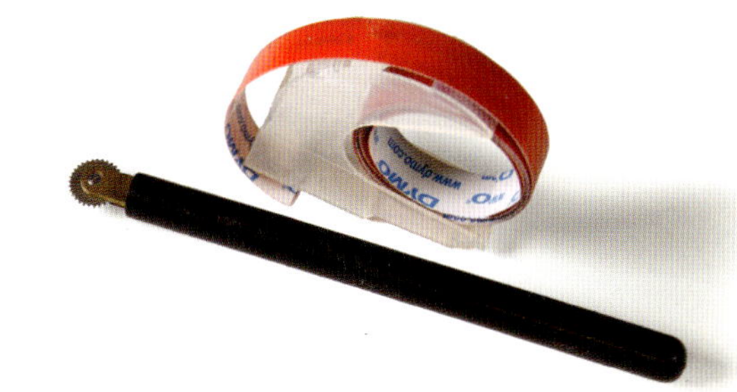

Some modern model kits represent stressed, corrugated, or damaged metal. However, many, if not most, don't. To re-create such textures, you'll have to up your scribing game. Take a breath, it's not that hard.

In general, you can use a hobby knife equipped with a No. 10 or No. 22 curved blade to scribe or indent the surface, shaving away just enough material to simulate the effect. Again, treat your work with some liquid plastic cement or fine sandpaper to soften sharp edges.

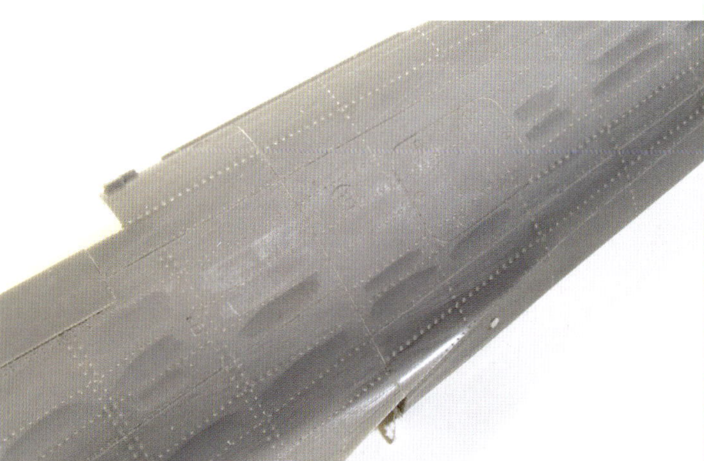

Assembling figures

Metal or resin figures are always broken down into separate parts. This allows for better detailed figures and means we can paint parts separately to reach areas that would be particularly difficult or impossible to reach on an assembled figure. Once you finish cleaning and prepping the parts, you can start putting them together. Large-scale figures often come with large locators that give the model strength and help ensure it's posed correctly. Smaller scale figures sometimes don't offer such a luxury.

Figure with large locators

Figure without locators

A figure without locators, or those with only small ones, needs reinforcements at the glue joints. Without reinforcement, figures are susceptible to break at even the slightest touch, no matter how good the glue. To avoid this, you should add brass rod to the joints, essentially making your own locators.

You can readily find brass rod in many diameters at your local hobby shop or online retailer.

1. For this procedure, you'll need cutting pliers and an electric or manual drill capable of grasping a micro drill bit the same diameter as your brass rod—1mm suits small scale figures; 3mm for larger scales.

2. Do this process for every joint. Before drilling the parts, make a small pilot hole with the tip of a sharp knife to give the drill bit a place to start and keep it from wandering.

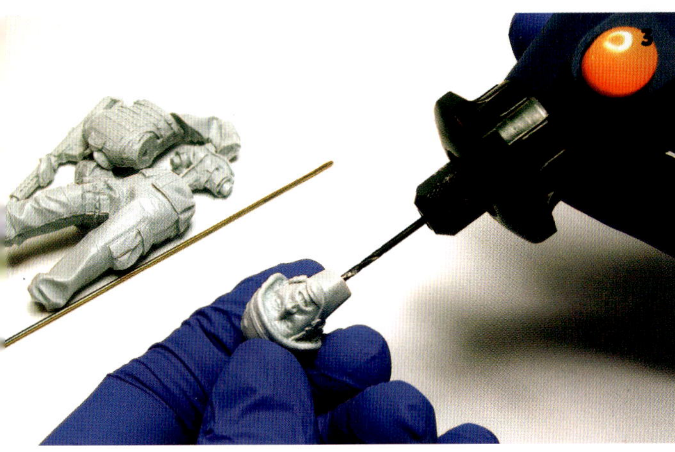

3. Drill deep enough to securely hold the rod, but no more. You can place a stop-mark with masking tape on the drill to ensure you don't drill too deep.

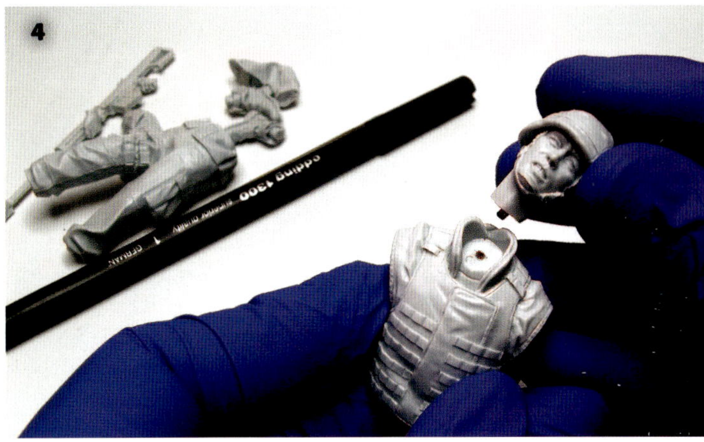

4. Mark the exact location of the hole on the corresponding part. Do this by inserting a toothpick into the hole you've already drilled. Apply paint to the end of the toothpick, then position the parts together. The toothpick will leave a mark where you need to drill. Dot the area with a permanent marker and wipe away the paint before it dries. Remember to remove the toothpick.

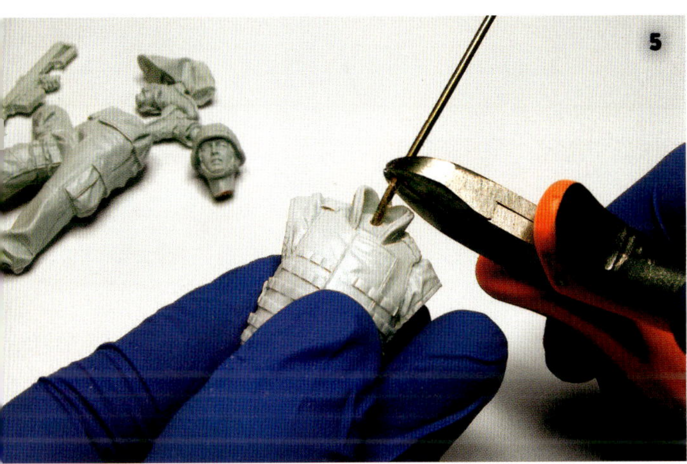

5. Trim the rod to fit the two parts together.

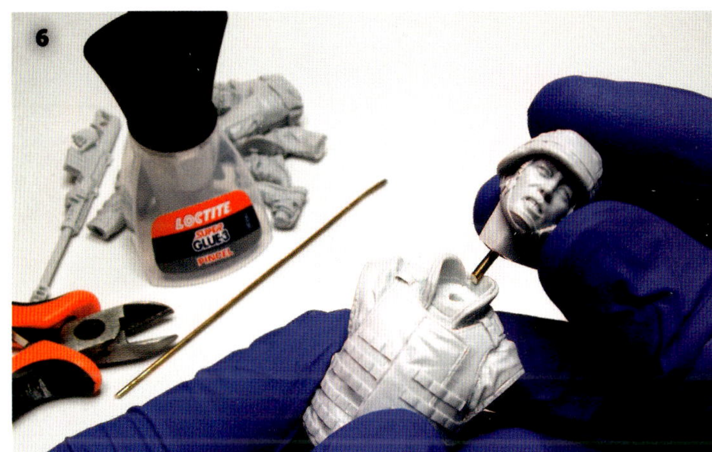

6. Test-fit the parts, make necessary adjustments, and glue both pieces to the rod and to each other with the appropriate glue.

Think ahead while building the figure. Will you be able to reach all the areas you want to if you assemble the entire model, or should you leave off an arm, leg, or a weapon?

Once the model has been assembled, fill gaps as normal.

TIPS

- If you're using an electric drill, use a low speed so you don't damage the part or injure yourself.

- Some figures are too small for reinforcement. Take extra care handling and finishing them.

- You can also use this technique to join broken or damaged parts, such as horse legs, hands to arms, feet to legs, and so forth.

Congratulations! You're about to paint your model.

Many novice modelers feel trepidation when faced with the prospect of painting and weathering. However, a grasp of basic painting fundamentals can make finishing a model surprisingly accessible and rewarding!

In this chapter, you'll learn about many of the techniques you can use to paint and weather your model. You'll also read about the products and materials available for finishing models, the proper ways to apply them, and the advantages and disadvantages each can provide.

Remember, this book is only a starting point. As your skills grow, you'll want to keep expanding your knowledge, researching and experimenting, and finding your own ways to accomplish particular tasks to achieve the desired results for your scale models.

3. Painting.

Color theory

Color theory provides a guide for how we can mix colors and what effects we can expect from color combinations. You should have a basic understanding of color theory to avoid selecting and mixing colors at random.

It's important not only to see a color, but also to distinguish it as either **additive** or **subtractive**. This categorization will help you correctly mix colors.

ADDITIVE and SUBTRACTIVE COLOR

Additive colors are immaterial, emitted from the sun or artificial light sources such as light bulbs, projectors, and flames. Subtractive colors have a substance and come from combining ingredients like dyes, inks, or paint pigments to create a wide range of material colors. These are the colors used for art and, of course, scale modeling. Subtractive colors are unavoidably affected by the additive colors from light sources.

The color black can only be produced by combining subtractive colors, while combining all the additive colors produces white. That means the combination of pigment colors (cyan, magenta, yellow) subtract the light, thus giving us black. The mixture of additive colors (fasmatic green, red, and blue) create light, so, white color is obtained. Light can drastically change the final appearance of your model. You must consider the model's scale and the light when choosing colors so your model doesn't look toy-like or historically inaccurate.

The next important thing to know is how to classify colors, no matter if additive or subtractive:

1

PRIMARY COLORS

1. Red, yellow, and blue are the primary colors. Primary colors can' be created by any combination of other colors. Instead, all other colors derive from these three. Primary colors in their pure forms must be used wisely, because they're dominant and attract the eye.

SECONDARY COLORS

2. Green, orange, and purple are the secondary colors. These colors can be formed by mixing equal parts of the appropriate primary colors.

2

TERTIARY COLORS

Tertiary colors are created when you mix a primary with a secondary color in equal parts. Such colors have a two-word name, like blue-green, red-violet, or yellow-orange. By mixing different quantities of different colors, we can create an endless palette commonly called the color wheel. The position of colors in the chromatic circle is not random and plays a significant role when it comes to combining them.

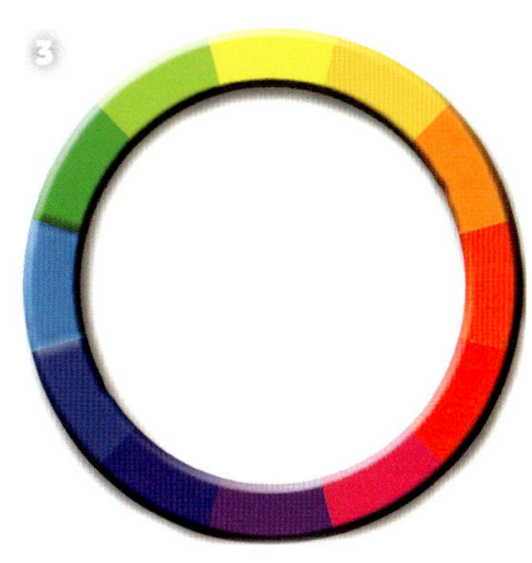

COLOR TEMPERATURE

What is the difference between and warm and cool colors? In the color wheel, there is an imaginary line that divides it into two equal parts according to the colors' temperature: Reds, yellows, oranges and their derivatives are warm colors, so noted because they supposedly create a feeling of energy and liveliness. Blues, violets, greens, and all colors originating from them, are called cool colors for their reported calming and relaxing properties.

ANALOGOUS COLORS

These colors come in groups of three and reside next to each other on the color wheel. They share a common color, with one that is dominant. Usually the dominant color is a primary or secondary color. Blue, green, and blue-green are analogous colors.

COMPLEMENTARY COLORS

Pairs of colors that cancel each other out by making a grayscale color are complementary colors. Traditionally, a complementary color pair includes a primary color and the secondary color opposite on the color wheel: blue and orange; red and green; yellow and purple. Complementary colors should be used carefully:

- The use of complementary colors side by side intensify each other.
- Complementary colors create contrast and can serve to draw the eye to areas of interest.
- Mixing complementary colors create gray, brown, earthy colors.
- Adding a small amount of a complementary color will reduce a color's intensity without changing the overall hue.
- To obtain the darkest shades of the colors, it's more appropriate to add the complementary color instead of using black.
- To get more striking areas of highlights, add a tiny bit of the complementary color in the white.

Color theory extends far beyond the information presented here, delving into light, hue, luminosity, saturation, and more. To start painting your model, it's enough to know the basics.

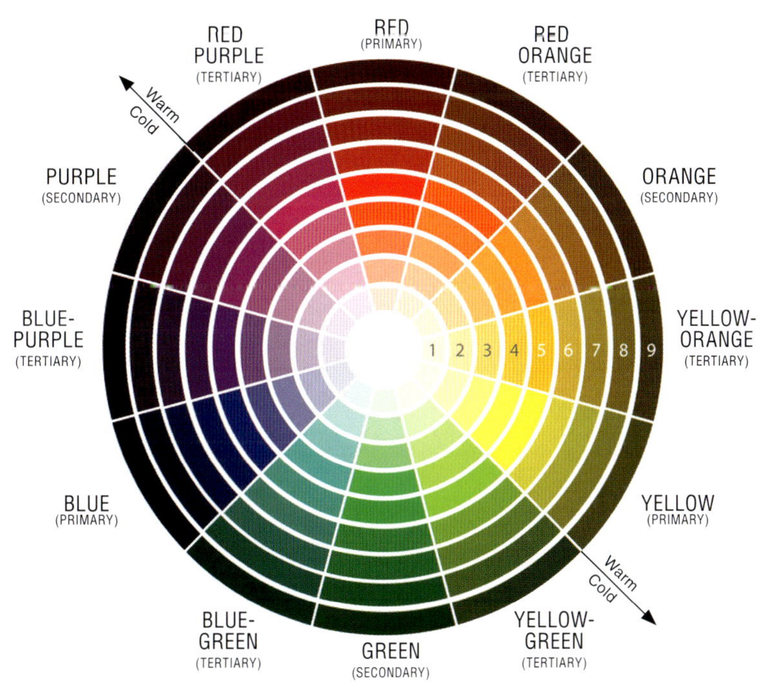

Color charts and scale effect

Entering the painting phase of a project, one faces the challenge of selecting colors that correlate to those on the real subject. Model paint makers have made great efforts to produce colors accurate to historic and contemporary subjects, making it relatively easy to make the right choices.

Well-known colors such as insignia red and German gray have been thoroughly researched and included in the Federal Standard color system, getting an FS prefix and number. Many times, the instructions for a kit will indicate the correct color to use on a particular part, often referring to FS colors. But you can look for additional information.

There are also color charts that show the equivalences of Federal Standard colors to colors offered by different paint brands, making it even easier not only to select the correct color for your model, but also to pick the equivalent from a brand you prefer.

The effect of scale on paint schemes is an often overlooked topic. The larger a surface is, the more light it subtracts. According to color theory, the final optic result of a specific color applied to larger or smaller surfaces can shift.

The more sophisticated paint manufacturers claim to have taken scale effect into account when formulating their colors. While this may be the case, colors are mostly adjusted to common scales: 1/35 scale for military vehicles and 1/72 or 1/48 scale for aviation-related subjects.

A haunting question then arises: Should you compensate for scale effect when painting a model? To answer the question we must remember that models are scale copies of larger objects.

It becomes a matter of perspective. For example, a 1/35 scale model seen from 12 inches away should look like the real object seen from a distance of 35 feet. Likewise, a 1/48 or 1/72 scale model seen from 12 inches away should look as if the real object was 48 or 72 feet away, respectively.

What you see at any given moment provides you with a grasp of size, distance and color. Looking at objects at a great distance gives the illusion of lighter, faded colors. This effect is known as atmospheric perspective. Air isn't completely transparent. We see things through a filter. Objects seen from far away are not sharp and clear. Applying this fact to our models, the smaller the scale, the farther away an object seems to be. At small scales, such as 1/144, the atmospheric perspective would make a color almost unidentifiable.

At larger scales though, such as 1/24, the color would be ever-closer to the one on the 1/1 subject. This phenomenon is related to the intensity or the amount of gray in a color. The intensity of a color such as red, ranges from pure red on the most intense side to reddish-gray on the faded side. In theory, neutral gray should be added to decrease the intensity of a color. However, if you try to reduce the intensity of an already subtle color, such as a light blue, by using gray, it will affect the delicate balance of a the color and possibly alter its hue.

For instance, colors used for camouflage already tend toward a gray, low intensity hue. Adding more gray isn't the way to decrease such a color's intensity. If the real scale color already contains enough gray, adding white lightens the existing gray, subsequently reducing the intensity of the color.

Therefore, to reduce the intensity of colors, particularly those used for camouflage, you can use this quick rule of thumb:

for 1/32 and 1/35 scale: 93% base color; 7% gray or white
for 1/48 scale: 90% base color; 10% gray or white
for 1/72 and 1/76 scale: 85% base color; 15% gray or white
for HO (1/87) scale: 80% base color; 20% gray or white
for 1/144 scale: 77% base color; 23% gray or white
for 1/350 scale: 70% base color; 30% gray or white
for 1/700 scale: 50% base color; 50% gray or white

Using these guidelines, dark green at 1/1 scale is 100% dark green; at 1/35 scale use 93% dark green and 7% white (because it is a low-intensity color that already tends toward gray; at 1/72 scale use 85% dark green and 15% white; and so on.

Paint thinners

Model paint comes in plastic bottles, glass jars, or even aluminum cans. Depending on the type of the paint (acrylic, enamel, lacquer) and the manufacturer, they are offered in many different consistencies. While some manufacturers offer ready-to-paint colors for airbrushes, in most cases paint comes in a rather thick consistency, making it difficult to use without thinning. To thin paint, you'll use acrylic thinners and solvents.

You may think there's no difference between acrylic thinners and solvents. While you may get similar results, they work very differently from each other.

ACRYLIC THINNERS

Acrylic thinners are specifically designed for use with acrylic paint. They add volume to the paint, making it less dense, but don't break down the paint's binding agent. Rather, they blend with it. The more thinner added to a mix, the smaller the overall percentage of pigment, so the color will become more transparent, but it still allows the paint to coat evenly. Thinning acrylics with thinner are perfect for creating glazes and filters. Thinner is also crucial when painting with an airbrush, because it provides the ideal medium to dilute acrylic paint to reduce the chances of it clogging your airbrush.

SOLVENTS

As their name implies, solvents dissolve paint, breaks down the binder, and distribute the pigment evenly throughout the solution. Increasing the percentage of solvent in your paint will make it increasingly less paint-like and more like the solvent, until it finally becomes transparent. By introducing solvent to paint you can create washes and stains. When it comes to solvents, you have no shortage of options.

ENAMEL AND LACQUER THINNERS
Not to be confused with acrylic thinners, these types of thinners are solvents. Used in small, controlled amounts, it helps to thin a paint. But add too much and you'll dissolve the color. Do not use enamel thinner with lacquer paint or vice versa.

WATER
Easy to obtain and practically free. Its the most common solvent and can be used with acrylic paints and water-based enamels and varnishes. Water has strong surface tension, so take care to avoid the formation of drops and puddles of paint.

TURPENTINE
Made by distilling resin from live trees, turpentine works well with enamel paint and artist oils. It has an extremely strong scent and is highly flammable. Always use in a well-ventilated area.

MINERAL SPIRITS/WHITE SPIRIT
This petroleum derivative is the most commonly used solvent for enamels and lacquers. It's also excellent to dilute artist oils. It's fast-drying and can be safely used for cleaning airbrushes and paintbrushes due to its mild properties. Its scent is similar to turpentine; use in a well-ventilated area.

DENATURED/RUBBING ALCOHOL
Easy to get and usually inexpensive, denatured alcohol works well for dissolving acrylic paint. However, because it dries so quickly, be careful when using it to clean your airbrush—it can create clogs. You can also use it to thin acrylic paint, but remember it also decreases its drying time.

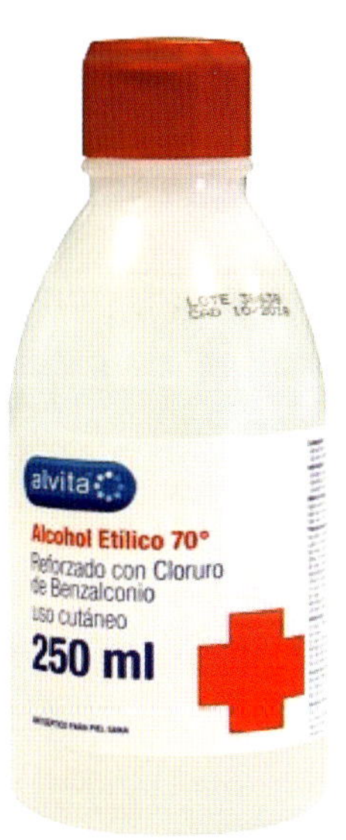

DISOLVENT UNIVERSAL
It's a very aggressive product based on toluene and recommended for dissolving of all types of paints and for cleaning tools. It's usually affordable, and can be purchased in large containers.

TIPS

- Glass cleaner (such as Windex) is a combination of water, alcohol, and ammonia. It's compatible with almost all acrylic paint. It has a fast drying time and low surface tension, which helps prevent puddling. It's a good solvent to have on hand.

- Always work with solvents in a well-ventilated area. Invest in a reusable respirator mask with organic vapor filters. Brushes used for cleaning with solvents should be used exclusively for the purpose.

Acrylic paint

Among all the paints that modelers use, acrylic paint is the most popular and easiest to find. Acrylics are composed of pigments suspended in a polymer emulsion. They dry quickly and, although they can be thinned with water, they are resistant to it when dry. When painting with acrylics, the results you can achieve range from a thick quality close to that of artist oils to thin and transparent, similar to watercolors.

Acrylic paints offer characteristics that can't be obtained with other types of paint. It's suitable for use with both a paintbrush and an airbrush, and is user-friendly, in both cases.

Although it can used straight from the bottle, it's advisable to thin acrylic paint with a purpose-made thinner or water, as this will provide better flow and a better finish.

Before using acrylic paint, vigorously shake the container to mix the pigment with the medium and get even color distribution.

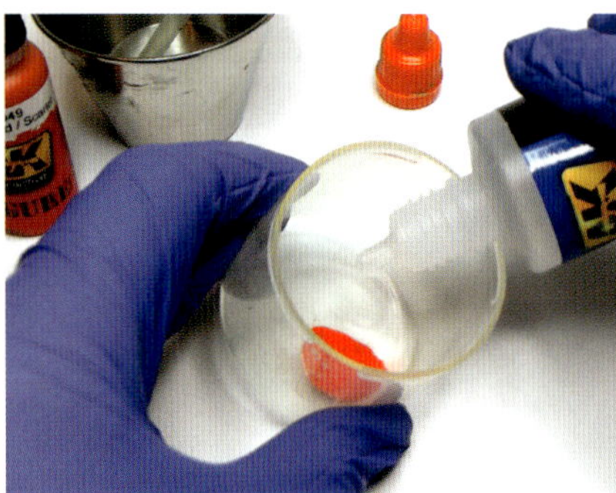

Pour the desired quantity into a container or a palette and add some thinner or water.

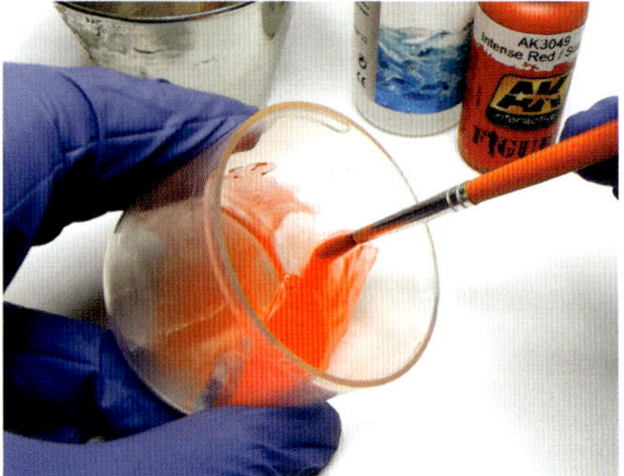

With the help of a brush, a mixing stick, or toothpick, stir the mixture. You can tell the paint is ready when it has a consistency similar to 2% milk.

Dip your brush in the paint and then press it against the inside of the mixing cup. If the paint drips freely down the side, then it's ready to go. If, instead, it flows slowly or runs quickly out of the brush, you'll have to add more thinner or more paint, respectively.

Varying how much you thin acrylic paint allows you to obtain different finishes. For example, using a highly diluted paint we can make a **glaze** or **filter** that will change the tone of the underlying color, without completely covering it. On the other hand, applying a less diluted paint, you can add body and even texture to the surface you're painting.

Acrylics have a short curing time, but sometimes you'll want your paint to stay wet or workable longer. In such cases, a specially made retardant can be added to your mix, always following the proportions indicated by the manufacturer.

Consider not only your personal preferences, but also the demands of the painting you're about to do when choosing between acrylic and other types of paint. Let's take a look at some of the advantages and disadvantages of acrylic paint.

ADVANTAGES

- Can be cleaned with water. Both paintbrushes and airbrushes can be cleaned with soap and water. Another option for cleaning is denatured alcohol.
- Fast-drying. Although the paint fully cures after approximately 24 hours, you can apply subsequent coats of paint within a few minutes. Acrylic paint retarders can give you more working time with the paint.
- The color chart currently offered in the market is extensive, so even exact colors can be readily found.
- Economically priced.
- Minimal toxicity.
- Versatile—use a paintbrush or an airbrush as necessary.
- Available in flat, gloss, and satin finishes.
- The best option for painting details and figures.

DISADVANTAGES

- Once dry, the resulting color can vary.
- It doesn't strictly respect the color theory. For instance, if you mix yellow and blue the result is not pure green, but rather a grayish blue. Experience and tests will help manage this issue.
- Although there are paints prepared for use with an airbrush, acrylics still require thinning. Typically, acrylic paint requires a 60:40 paint-to-thinner ratio. Use an acrylic paint retarder while airbrushing to avoid clogs.
- Acrylic metallic paint and weathering-effect mediums are usually less effective than enamel-based products.

A wet palette can be extremely useful when working with acrylic paint. A wet palette combines a sponge, palette paper and a tray to keep paint moist and workable, even for up to days at a time if you have a cover. You can find wet palettes available online or at hobby or art supply stores. You can also easily make one yourself.

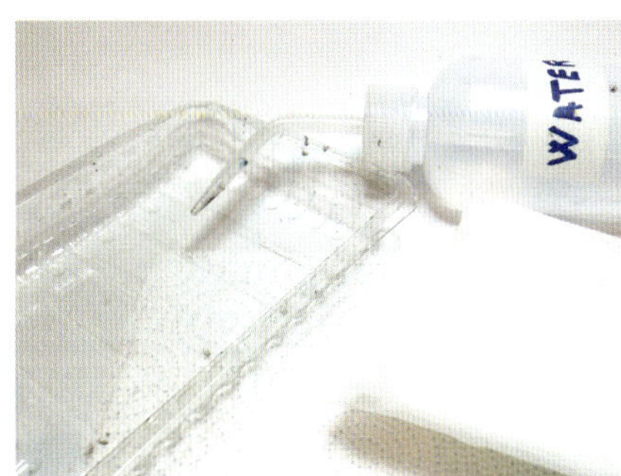

To make your wet palette, you'll need a container, paper towel, some parchment paper, and water.

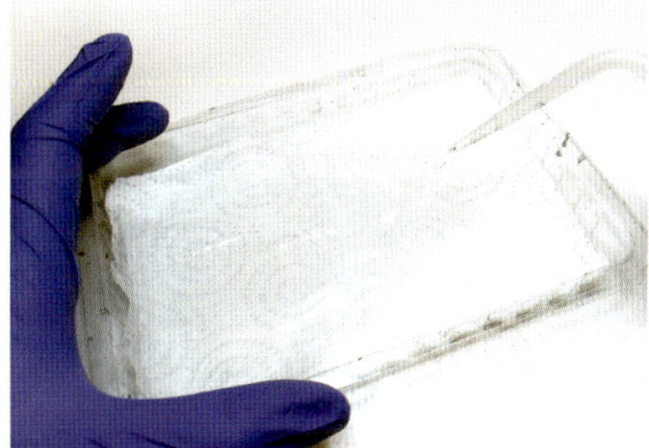

Cut the paper towel to fit inside the container and cover the entire bottom. Add three or four layers and soak them with water.

Cut a piece of parchment paper to fit into the palette, and place it on top of the paper towel. Your wet palette is ready to go!

TIPS

- Keep a variety of small cups, dishes or other containers for mixing paint.

- Never mix the paint directly in the airbrush; you might create a clog. Although nontoxic, always airbrush acrylic paint in a well-ventilated area and wear a protective mask.

Lacquer paint

The general term lacquer paint sometimes generates confusion and questions about the paint's real composition, mainly because some manufacturers call it acrylic lacquer and others simply lacquer. They all mean the same thing.

Solvent-based lacquer paints are composed of an acrylic resin dissolved in a nonaqueous medium, making them generally incompatible with water. There are some water-based lacquers on the market that can be used with water, but they can be reactive to other types of finishes. In general, lacquer paint is thinner than acrylic paint, is less likely to clog your airbrush, and has good color density when applied.

Many modelers make extensive use of lacquer, and several brands offer a wide range of colors. Their popularity is largely attributed to the hard and durable base they provide once dry, easily enduring even aggressive finishing processes. Military vehicles of any scale and time, airplanes, ships and also civil and science fiction subjects look perfect with a durable and versatile lacquer finish. It is without doubt one of the best paint options.

ADVANTAGES

- Airbrushes well
- Durable finish
- Extensive ranges of colors
- Smooth, glassy gloss finish; perfect flat

DISADVANTAGES

- Must be sprayed from an airbrush or aerosol can; airbrush for optimum results
- Avoiding a respirator mask isn't an option
- Airbrush must be cleaned thoroughly between applying different colors; constant maintenance slows progress and becomes tedious

How to use lacquer paints

Start by stirring the paint well in its bottle with a mixing stick. Make sure to mix the pigment evenly, reaching all the way to the bottom of the bottle and to the corners. After the color and consistency of the paint are even, you can decant a portion into another mixing container. Thin your lacquer with an appropriate solvent or name-brand thinner. Stir the paint with your mixing stick until it has the consistency of 2% milk and the color is even throughout and then pour it into the airbrush.

While not always the best option, you can pour an amount of thinner into the airbrush color cup with the help of a pipette. Note: This should not be done with acrylic paint.

Using a clean brush, add the lacquer paint, and stir with a toothpick until you get an even consistency. It's always preferable to paint with a highly diluted paint rather than with an under-diluted one. Pure paint can cause occlusions in the airbrush and produce an **orange peel** texture on the model that will have to be removed and refinished.

Apply thin and even coats in successive layers. Move slowly and deliberately and don't focus on a single area—that's how you get runs in the finish and obscure fine details.

Lacquer is likely the most popular paint for finishing cars, airplanes, ships, and other vehicles. It's seldom used for figures except for perhaps a base coat.

TIPS

- Wear a respirator mask and paint in a well-ventilated room.
- Always apply in thin layers. Don't try to cover the model in a single coat of paint.
- Use the correct thinner or you can ruin your model and damage your airbrush.

Enamel paint

For decades, enamels were basically the only choice available for scale modeling. By and large, enamel paint is made of pigment mixed with an organic-oil base, although in recent times there have been more latex- and water-based enamels on the market. To work with enamels effectively, you must thin them precisely with the correct type of thinner.

Enamel paints take longer to dry than any other paint. Purpose-made additives can be used to shorten the curing time. Enamel comes in gloss, satin, and flat finishes. Due to its oil base, enamels flow and adhere well, making it the preferred medium for shading and weathering effects. Of course, they can be used for general painting too.

ADVANTAGES

- Large variety of affordable paint colors and finishing products
- Don't change color when dry
- Ideal for mixing and blending because of long drying time
- Best coverage properties of any paint
- Economical
- Can be applied with an airbrush or by hand

DISADVANTAGES

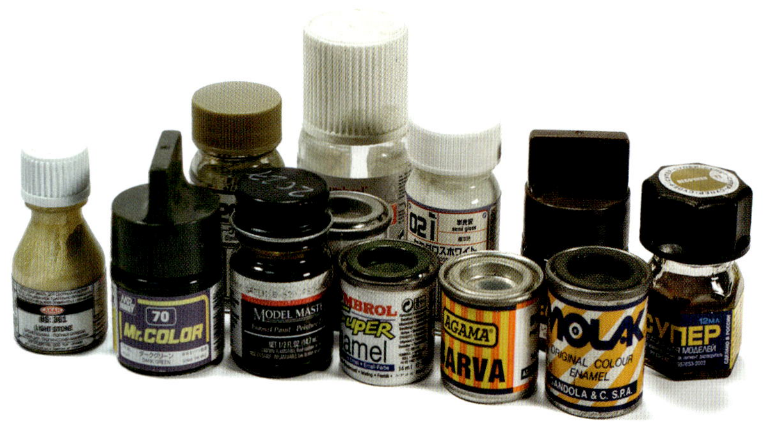

- Long drying time—between six and 24 hours or up to several days depending on humidity and temperature
- Solvents and thinners for enamels are aggressive, toxic, and have a strong odor
- Cleaning paintbrushes and airbrushes can be more complex and time-consuming due to enamel paint's strong adhesive properties. Take special care when cleaning your airbrush because enamel residue can become rock-hard and cause persistent clogs in your airbrush

If you want to obtain a solid color, using enamels is very simple. First, dilute some paint with the appropriate thinner, and then paint your model either with an airbrush or by hand. Remember the long drying times involved.

Working with enamels can be complicated if we want to blend different colors to obtain effects. For example, let's look at how you might go about painting groundwork, creating areas of shade and light similar to what you would see in nature. You'll need an appropriate thinner, enamel paints, and containers in which to pour your custom colors. Use a brush with synthetic bristles to apply enamels. Synthetic bristles are more durable and can better withstand the strength of enamels and their aggressive thinners.

Begin with a base coat. This first layer must always be complete because all the layers to follow will be semi-transparent.

With the base coat down, proceed by applying different shades, both lighter and darker, to simulate light effects and give life to your terrain. Shadows can be painted by adding black to the base color, while highlights can be created by adding yellow or even pure white to the base color.

Heavily diluting black, you can paint dark shadows.

A filter of the base color applied to the edge of the shadows creates a realistic transition from dark to light.

With the two colors still wet, use a brush to blend both tones, thus creating a pleasant and realistic gradient.

Due to its slow drying time, enamel paint even allows you to work directly on your model without using a mixing palette. This offers the advantage of continuously adding colors and filters until the desired effect is achieved. In the photo, white was added a little at a time and blended in with the rest of colors until a perfect gradient effect was created.

For more contrast, you can mix other colors into the shadows, like blue to give a sensation of cold or wet, or red tones in the gradient on top to make the results pop. Greens can simulate moss deposits.

TIPS

- Always use enamels in a well-ventilated place. The vapors are toxic.
- While enamels can be used on figures, consider using acrylic paint instead.

Spray paint

Spray paint often brings to mind graffiti or a means of quickly painting something around the house. However, spray paint can be very useful in a number of aspects in scale modeling.

Pressurized spray-paint cans contain the ready-to-paint color mixed with a gas propellant. While spray paint can speed up the painting process, it's not accurate like an airbrush or a paintbrush. Spray paint is perfect for applying base colors, primers, or varnishes.

Spray paint can be divided into three categories:
- Paints: colors used to paint our models
- Primers: purpose-made paints with extra adhesion for base coats
- Varnishes: transparent finishes that seal and protect

Spray paints have unique characteristics. The composition of the aerosol mixture has a strong and potentially toxic odor, so spray-painting should be done outside or in a place with a ventilator fan. Protective gloves and a dust mask are recommended.

Paint exits the pressurized container at high speed and volume, so you can't avoid overspray. And a cloud of paint particles typically forms that can attach to objects you don't want painted. Lastly, for the best results with spray paint, the surface of the model must be clean, sanded, and smooth.

Always use a primer spray first to prepare the surface for the final color. Apply it in thin coats and gradually build up the color.

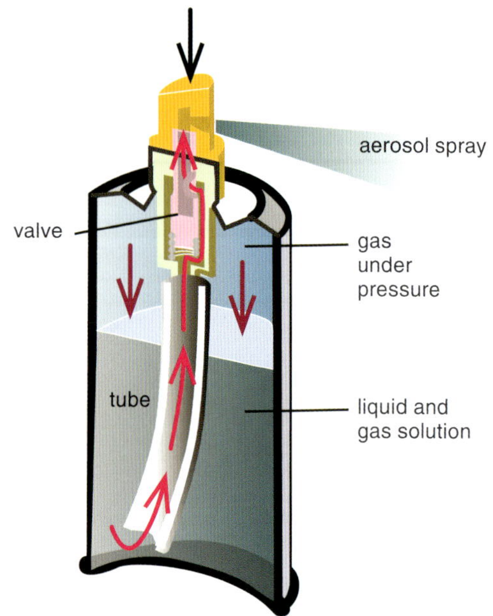

Before using any type of spray paint, shake the can for 1 to 2 minutes. Test spray the can away from the model to check the pressure and make sure paint is coming out in a spray, not spurting or dribbling. Apply the paint, keeping the nozzle about 12 inches away from the model's surface. Hold the can vertically and press the valve and pass the paint over the model with gentle movements, beginning and ending the pass clear of the model.

While painting, you'll notice the spray can gets cold and the paint exits at a reduced rate. To ensure the aerosol pressure remains constant during application, keep the paint can at 60°F (16°C). Immerse about three-quarters of the can in warm water for a few minutes before painting. **DO NOT** introduce a source of direct heat, such as a flame or stovetop heating element, to a pressurized canister.

You can clean up spray paint, depending upon its type, with a solvent or soap and water. The valve and nozzle should be thoroughly cleaned after you've finished. Invert the can and press the valve until you're only expelling propellant and no paint. This clears out any remaining residue and avoids clogs in the valve and nozzle.

Spray-paint cans lose pressure with each use as the propellant gradually runs out. There's nothing you can do about it. However, be careful because low pressure can cause spurts and uneven coats of paint.

As a can begins to lose its pressure, one option is to decant the paint and use it in your airbrush instead.

To transfer spray paint from the can to your airbrush, carefully spray the paint into a container. To help with this process and cut down on the mess, securely tape a straw to the nozzle. Some spray paints come with a special nozzle designed to decant paint.

SPRAY-PAINT NOZZLE SETS

Painting with spray-cans be effective and quick, offering smooth and durable finishes. Yet, it lacks the control of an airbrush. A good way to adapt the valve and nozzle to our painting needs is to buy a spray-paint nozzle set (or cap set).

Nozzle sets can come with specialty nozzles that create different spray patterns from thin to wide, and change the orientation of the spray from horizontal to vertical and all points in between. Some specialty nozzles can even control the pressure with which the paint leaves the can, offering even greater accuracy.

Artist oils

Artist oils, or oil paint, are a type of paint usually associated with fine art. However, they can be effectively used in scale modeling, too. Composed of mineral pigments mixed with an oil (traditionally linseed oil) that dries when exposed to air, artist oils are water insoluble and chemically inert.

The resulting mix is a thick, creamy, paste almost exclusively packed in flexible tubes. Like enamels, artist oils can be mixed and thinned with turpentine, mineral spirits, or any other organic solvent. Artist oils, even when heavily diluted, are always applied with a brush, because their thick consistency makes their use with an airbrush impractical. Keep a set of paintbrushes exclusively for use with artist oils in order to avoid contaminating them with residue from other types of paint.

In the scale modeling world, artist oils are still widely used for figure painting. They provide excellent blending properties and allow perfect transitions and gradients. Even so, many modelers have replaced them with acrylics due to the comfort and ease of use the latter provides. Apart from figure painting, artist oils can be also used for many effects on models from weathering and washes to filters and glazes.

ADVANTAGES

- Excellent color fidelity that strictly conforms to color theory
- Highly versatile; when very thin, artist oils make great washes or filters. Artist oils can be thickened using raw pigment or an oil-based paste to provide more volume for textured finishes (impasto)
- Durable medium that retains its color over time
- Can be applied over a variety of surfaces without fear of damaging the underlying material

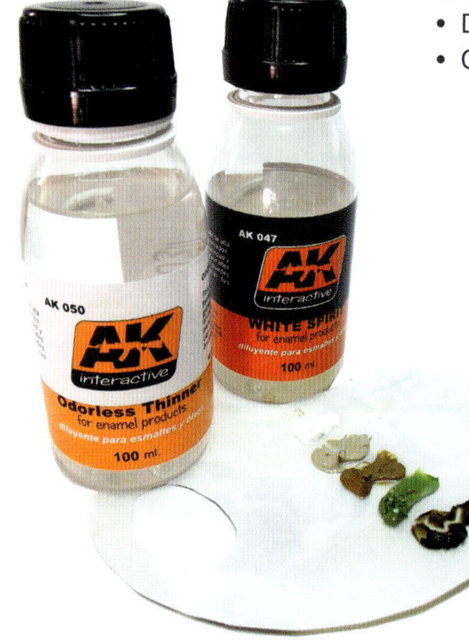

DISADVANTAGES

- Because artist oils are diluted with turpentine, which is toxic and has a strong odor, they must be used in a well-ventilated room.

• Require a long time to dry. The drying period can be reduced or lengthened using purpose-made additives
• Difficult to clean after it has been applied
• Need a palette or other dedicated working surface to properly mix and paint

When working with artist oils, you must always use a palette to mix the paint. There is no need to use a wet palette, as there is with acrylics, because the drying time for artist oils is much longer, leaving us plenty of time to work. There are many different ways to use artist oils, depending to the result you're looking for.

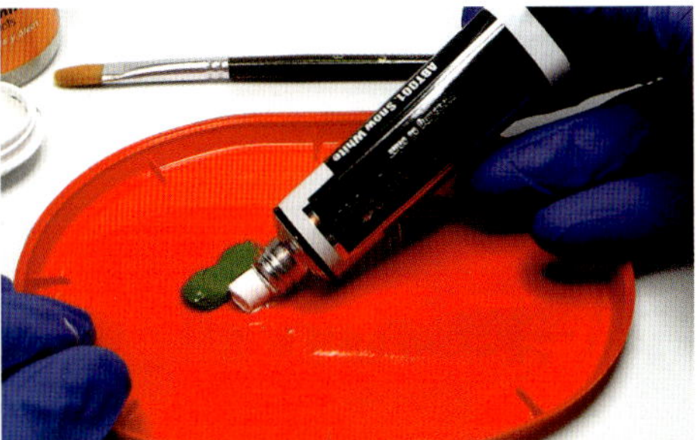

Place all the selected colors on your palette.

With the help of a brush and thinner, mix the paint until you achieve a smooth transition between the colors.

The more solvent used in the mix, the thinner the result. If you want to add texture or volume to a surface, use little or no solvent at all.

With a 9:1 mineral spirits-to-paint mix, you can use artist oils to make filters, glazes, and washes.

Mix the oil with the right proportion of mineral spirits.

Apply the wash in those areas where you want the color to accumulate.

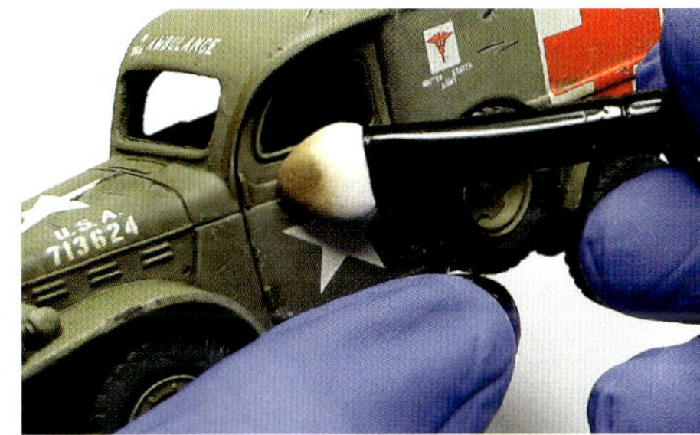

After waiting 10 to 15 minutes, remove any excess paint with a make-up remover or cotton swab.

Another way to work with artist oils is to mix and blend the colors directly on the model. Apply the oils to the surface, and with the help of a brush impregnated with mineral spirits, blend them together. This technique can add both color and texture to your model.

TIPS

- By mixing and blending colors you'll get the tones and shades you're looking for. Don't be afraid to experiment.

- Keep two separate containers of thinner: one for the paint and another for cleaning brushes.

Pigments

Pigments are basically the powdered material used to manufacture colors. Pigments are derived from pulverizing various natural minerals, such as magnesium, titanium, and zinc. The quality of a pigment depends on two factors: purity and the grade of powder—the finer the powder, the better its quality.

With so many brands and pigment colors available, we can select the best ones for our needs. In scale modeling, pigments are mainly used for weathering and wear effects. Pigments can effectively simulate dust, dirt, and mud. While you can apply pigments in their raw form, they lack any kind of binder and depend on friction to stay in place. To permanently affix them to a model, you use can use purpose-made pigment fixer, denatured alcohol, or acrylic thinner.

DRY PIGMENTS

The most common method for using pigments is to apply the dry powder onto the model's surface with a soft brush. To avoid losing your pigment effects over time, apply a small amount of pigment fixer onto the powder.

On the piece of groundwork below, you can see an area with exposed dirt and vegetation. There are also some concrete elements. Earth tones for the ground and grays for the concrete can really dress it up, adding realism to the piece.

Apply lighter pigment to the most illuminated areas and darker pigment to the shadowed areas. You can mix pigments by combining two or more for a wide range of shades and effects. Brush the pigments on the bare dirt areas, including the stones and the magazine—all the elements have to blend with the surroundings. Take care not to create large accumulations of pigment creating an unnatural appearance.

To keep the pigment in place, soak a clean brush in pigment fixer. Then lightly touch the brush to the surface. The fixer spreads into the pigment through capillary action. Never brush the fixer on; doing so would cause the end result to turn muddy. Work methodically, concentrating on small sections until you've applied fixer to the entire model, and let it dry.

WET PIGMENTS

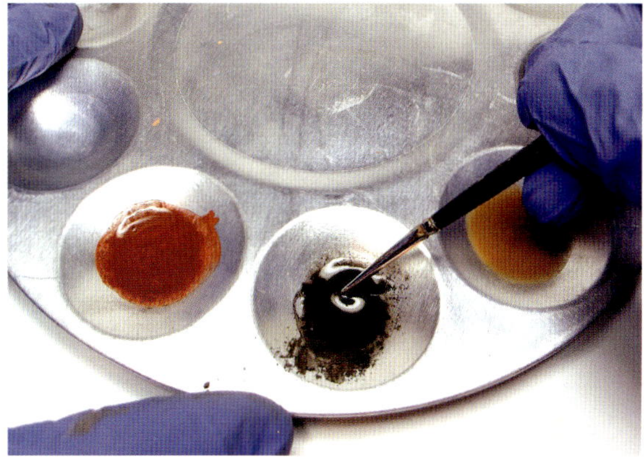

Wet pigments open weathering possibilities you can't achieve with dry pigments alone, simulating grease, smoke stains, rust, dirt streaks, even mud spatter. Place a small amount of pigment in a mixing cup or on a palette and add a few drops of fixer. Mix them until a liquid, translucent color is created. Depending upon what you want to use it for, a thicker mix will allow you to create texture; a more liquid mix will let you flow the pigment into gaps and along corners.

Brush the pigments in areas where you want to weather. The more you concentrate on a particular area, the dirtier it will look. Rather than try to get all the weathering right the first time, give your work time to dry. Find those places you'd like to see more weathered and repeat the application. Do this as many times as necessary to get the desired effect, rather than overdo it with a heavy hand.

TIPS

- Adding pigments is usually the last weathering step in a build, so it's always preferable to have your model varnished first.
- Pour an amount of dry pigment onto a working surface rather than using it directly from the jar. There's less chance of contaminating the color.

Pastels

The use of pastels has largely been forgotten and neglected by modern modelers. Yet, it's interesting to know this material and its properties.

Pastels are colored sticks resembling chalk both in look and consistency. They are used mostly by artists and can be found in art stores. Although modelers often try not to use materials that aren't designed and marketed specifically for the hobby, almost any painting and finishing material can prove useful in time. Pastel chalks can be scratched with a razor or rubbed on sandpaper to provide a very fine colored powder. The powder can be applied with a brush to create delicate weathering effects. Pastels can even replace some effects usually made with an airbrush!

Apart from the common chalk-like form, pastels can also be found as pencils—a feature very useful when it comes to fine detailing. The most useful pastel colors are black, white, yellowish ocher, burnt sienna, burnt brown, and dark brown. This array is enough to derive many different shades and tones if properly mixed. When applying sandpaper-powdered chalks, it's advisable to use cheap brushes, but with soft hair. Rubbing the brush on sandpaper to accumulate the powder eventually destroys the bristles.

DIRT AND WEATHERING

First, decide what colors are suitable for the model you're weathering, or simply choose the color of dirt you want to simulate. If there's no pastel chalk available of that color, you can always mix different tones to get what you need.

Load the tip of your brush with the pastels and make long brush strokes. You can see how the pastels change the tone of the model's colors. Blend the pastels with a soft fan brush.

Pastels, like pigments, can't permanently adhere to a surface on their own. If left unfixed, pastels will quickly fall off whenever the model is handled or dusted. The best way to fix pastels is under a protective layer of varnish. An airbrush makes the perfect tool for this because you can moderate the air pressure and better control your aim so you don't blow away the pastels you've worked so hard on.

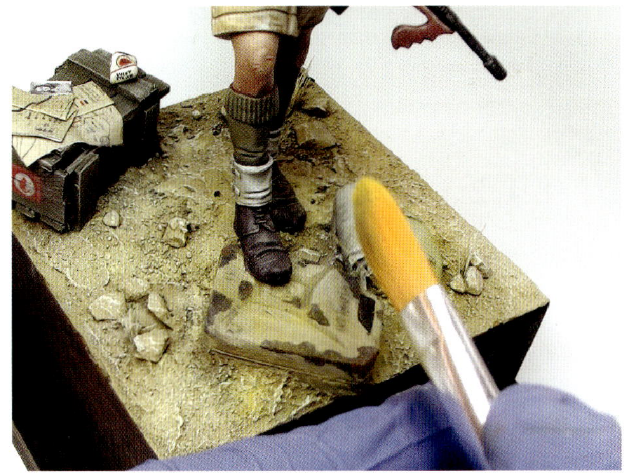

Another option is to apply pastels by gently hitting a brush loaded with powder against your finger, keeping the brush a short distance above the surface. The pastel dust will fall onto the surface, imitating the effect of splashes.

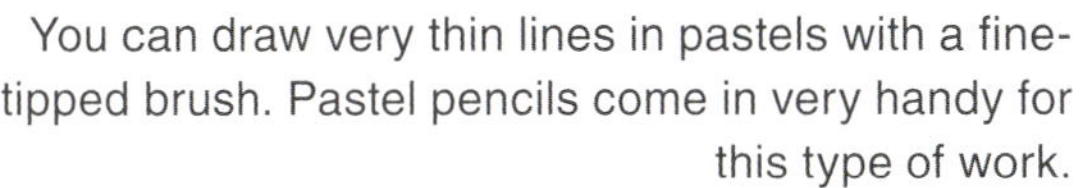

You can draw very thin lines in pastels with a fine-tipped brush. Pastel pencils come in very handy for this type of work.

Pastels can easily imitate dirt. Again, load your brush, tap off the pastels, then gently blow on the powder, letting the natural air currents do the work.

It may seem that this material is perfect to weather you model. Pastels are inexpensive, easy to apply, and provide great results. However, pastels are extremely fragile. Pastel sticks frequently break, and they must be kept separate from other pastels so they don't stain the other colors. Finally, purpose-made fixers for pastels, often used in fine art, don't work as well for scale models because they tend to darken and distort the original colors.

TIPS

- Using pastels on a model must be the last step because any subsequent manipulation of the model will almost certainly ruin the results.

- Alcohol can be used as a fixer for pastels, but the process should be tested first because the results can vary widely, changing the pastels' color and consistency, as well as affecting underlying finishes.

Metallic paints

Imitating metal realistically on models has been a huge headache for modelers over the years. The best way to achieve a metallic look on our project is to use paints that contain particles of metal in their composition. Such paints provide the natural shine and color of metal. But using these paints can be more complicated than it appears.

METALLIC ACRYLICS

Acrylic metallic paint are mostly used on figures or small details on larger models. You can brush them on by hand or with an airbrush. As with other acrylics, they can be thinned with acrylic thinner or water. It's important to apply acrylic metallic paints well diluted, since their composition is usually thicker than other acrylics. If not properly thinned, they will give an undesirably rough finish. You can find a wide range of metallic acrylic paints from many manufacturers, and they can be intermixed with each other or nonmetallic acrylics to get the desired finish.

STEEL
There are many different shades of steel. For a basic steel, begin with a dark color as base—steel mixed with black or dark blue. Gradually add lighter tones to create highlights and provide shine. The areas or spots of maximum highlights can be painted with a polished steel color.

BRONZE
Work the same way with bronze as you did with steel. Darken the bronze with black or dark green to paint the shadow areas. Add more bronze to lighten for mid-tones. In areas of extreme shine, mix bronze with polished steel.

GOLD
Add dark red to gold for shadow areas. Gradually add highlights by adding more gold to the mix. The finish is completed adding final highlights with polished gold or polished steel to the gold.

METALLIC LACQUERS

Developed with high-quality pigments and specially formulated to provide optimum results, metallic lacquers can make for hyper-realistic finishes. They are very durable and resistant to masking and weathering, and create wonderfully even coats. Metallic lacquers cure almost instantaneously, but they make cleaning airbrushes a chore.

Start with a black base coat. This will help provide a mirror effect. Next, with smooth, even motions, airbrush the lacquer over the entire surface.

Bear in mind that models with complex surfaces can cause the metallic sheen to appear inconsistent. To overcome this problem, airbrush more layers of lacquer on the desired areas after the first one dries. Give each successive coat an hour to cure before painting the next one.

When finished painting, polish the surface with a soft cotton cloth to increase the shine. After 24 hours, you can begin weathering.

METALLIC WAXES

Metallic waxes are created with high quality metallic pigments incorporated into a wax-paste base. Waxes can supply durable and very realistic metallic effects. When dry, they can be polished to an incredible shine. Given its clunky application, is not recommended for small details, but it can be used on any type of model. Just spread the wax on a surface with a brush, cotton swab, or even your fingertip, and let it dry.

After 24 hours, you can polish metallic wax with a cotton cloth until you get the desired shine.

If a spot needs more wax, brush on some more and repeat the polishing process until you have a uniform finish.

When you've finished polishing you can move on to weathering without worrying about the metallic wax.

Varnish

The main purpose of a varnish (also called clear coat) is to protect and preserve your model from dust and humidity. Varnish also provides a durable base to withstand aggressive finishing processes.

You can find acrylic, enamel, and lacquer varnishes, ranging from flat or matte (no shine) to satin (some shine) to gloss (high shine). In addition to the standard finishes the manufacturers provide us straight from the bottle, a wide variety of different finishes can be obtained by mixing varnishes. For instance, the sheen of gloss varnish can be reduced by adding some satin or flat varnish. Test and experiment to get the desired look. Moreover, you can have different finishes on separate parts of the same model, thus enhancing the sense of different material or textures. Add a few drops of varnish to paint color to creatively alter its finish.

Before airbrushing any varnish, make sure the model is clean of dust, hair, or other dirt that you don't want captured in the finish. You can brush on varnish by hand, but unless it's an ultra-flat finish, use an airbrush. Gloss varnishes are particularly hard to brush by hand and get an even coat without brush strokes.

Airbrush varnish on in light, uniform layers, until you achieve a smooth and homogeneous finish.

You can see the differences between gloss, satin, and flat varnishes.

While clear coat is often applied at the end of a build, that isn't always the case. Sometimes you have to clear coat in the middle of a build because trying to apply it at the end would be too difficult and risk ruining another part of the model.

• When you paint scale figures, especially if artist oils have been used, the finish tends to be satin. Correct the sheen with flat clear coat. Metallic paint should be finished with gloss varnish. Some modelers use gloss varnish on the eyes in large-scale figures.

• Varnish fully integrates decals on a model's surface. First, the area has to be covered with a layer of gloss. After you've placed the decals and they've dried, airbrush a second layer of varnish over the decals to create a shell to protect them from possible damage during weathering.

• Military vehicles can be painted either with flat or satin varnish. Civilian vehicles are usually gloss.

• Use varnish between finishing layers. For example, apply varnish after you've finished with a wash to lock in the work you've done. Then continue with the next weathering process.

• Gloss varnish can simulate wet surfaces, such as sweat, saliva, and rainwater.

A coat of flat varnish can sometimes eliminate or at least minimize the effect of decal silvering. But remember to apply decals to a gloss surface for best results

Varnish, while wet, can be used to glue small details such as leaves, sand, and gravel into place.

TIP

- Variety is something we should always look for, so don't use a final and general coat of varnish on a model. In the real life, most objects have both flat and gloss parts.

Paintbrushes

Paintbrushes form the backbone of any modeler's painting and finishing tools. There are some modelers so skilled with brushes they forgo airbrushes—rare, but they exist. When you're just starting out, it can be hard to know what types, sizes, and shapes of paintbrush you should buy. With so many available in so many different price ranges, which are right for you?

Because you will probably work with some pretty aggressive finishing products, you can steer clear of the high-end fine-art brushes. On the other hand, don't buy the cheap ones either. Mid-range paintbrushes with synthetic bristles that are soft but retain their shape are perfect for most applications. For figure painting, edge more toward the expensive end; Marta Kolinsky paintbrushes are a good option.

At its most basic, a paintbrush consists of an elongated piece of wood or plastic that, at one end, has a plume of hair, called **bristles**, held in place by a piece of metal called a **ferrule**.

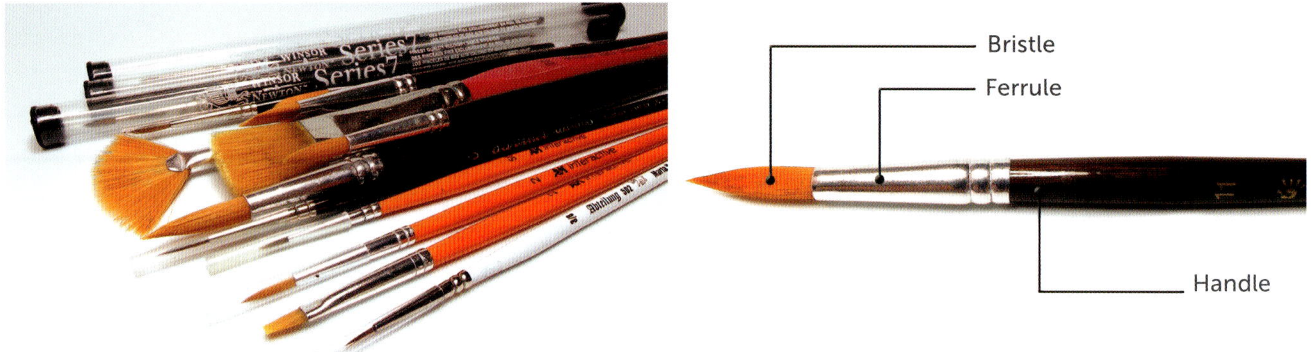

There are two categories of paintbrush:

ANIMAL HAIR

Animal hair paintbrush bristles come in a wide variety of types, including badger, camel, Kolinksy sable (actually mink), and pony hair. Each hair type has paints that it can be used with and others it can't. For instance, badger hair brushes are used exclusively with artist oils; pony hair brushes are more versatile, good with acrylic, tempera, and watercolor.

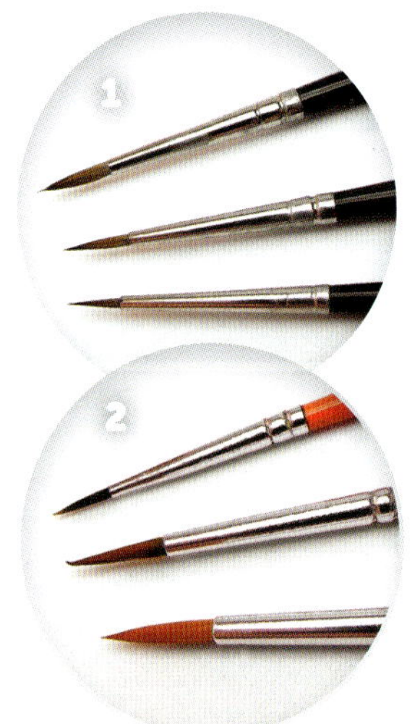

SYNTHETIC HAIR

Made with nylon or polyester fibers, synthetic hair brushes tend to be more resistant to damage from solvents and paints. They're easier to clean than animal hair brushes and far more durable. Because of their resistance to caustic paint products, synthetic brushes are the popular choice for painting with acrylics, but can be used with any paint.

Next, you must choose the shapes of the paintbrushes you'll need. Here are the most common:

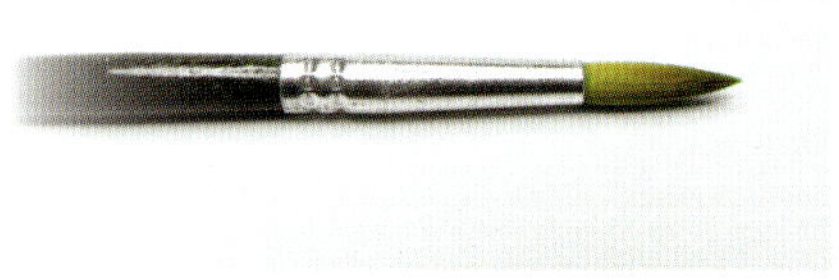

ROUND BRUSHES

Round ferrule and pointed-tip bristles. Ideal for painting details and lines.

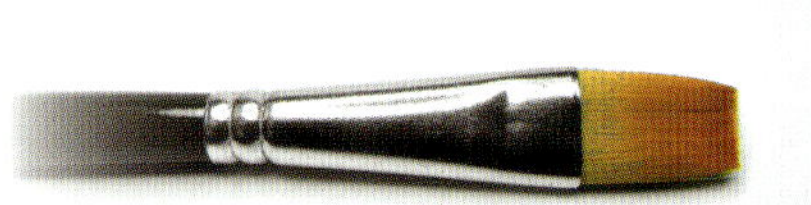

FLAT BRUSHES

Flat ferrule and broad hair-tip. Ideal for painting large surfaces.

ANGLED BRUSHES

Same as flat, but with an angled tip. Great for reaching difficult and remote spots with enough paint to cover it.

FILBERT

Flat ferrule; thick, rounded tip. An excellent brush for dry-brushing and brushing on pigments.

FAN

Flat ferrule; soft, fan-shaped tip. Use to apply and blend pigments.

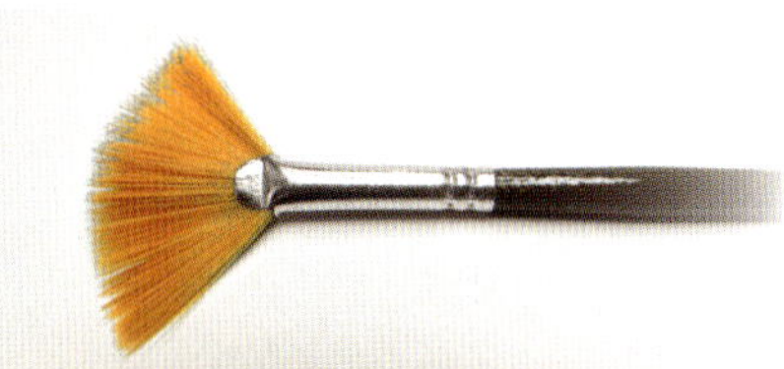

COLOR SHAPER TOOLS

Various shaped tips. Not brushes per se, they can be used to apply acrylic paint or artist oils straight from the bottle or tube and then carve into the paint for different effects.

The size of the paintbrushes you use depends entirely on the sort of painting you'll do. The most common modeling sizes for round-tip brushes are 0, 1, and 2 (the lower the number, the smaller the brush).

While you can produce an array of effects with a single brush by varying your painting technique, it's better to have paintbrushes specifically for certain tasks. For example, while you can use a fine-tipped brush to paint both delicate and wide lines, they really are suited to the former. Better to have a larger brush, or even a flat brush when you need to cover a large area.

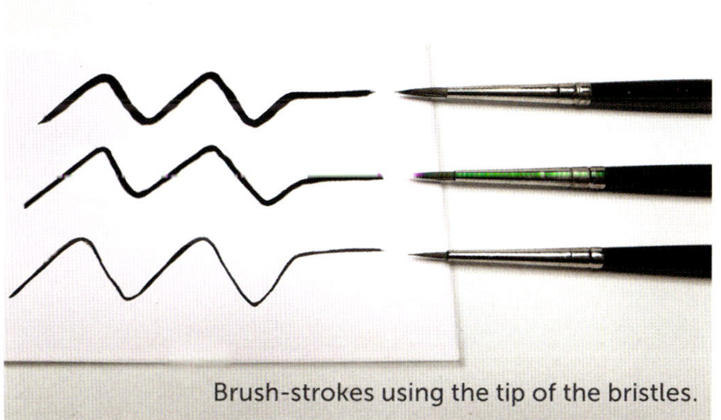

Brush-strokes using the tip of the bristles.

Brush-strokes using the whole bristle area.

TIPS

- Protect the tip of a paintbrush with the plastic sleeve it came with and store it with the bristles standing up.

- Clean your brushes well after each session and use brush preserver to lengthen their life span.

- Keep your old brushes; use them on procedures you know tend to destroy brushes. Dedicate brushes for particular tasks, like weathering with pigments and dry-brushing, and use them exclusively for those tasks.

Using a paintbrush

Seeing the work of experienced modelers, one can wonder how they get such a perfect and clean finish. You may initially think it's achieved, apart from their experience, by the use of an airbrush, which offers an even result with smooth transitions. This is not always true, although experienced modelers do combine airbrushed and hand-painted techniques. Certain effects can only be done with a paintbrush. An airbrush is an expensive tool, and not all modelers have one. However, all modelers have paintbrushes on their workbenches, no matter the level of experience.

While it seems obvious, the first thing you need to learn is how to hold a paintbrush, so you can work with it without getting fatigued. While many models can be secured on supports or placed directly on your workbench, you can hold the model in one hand and paint with the other. Place your elbows or forearms on the workbench to reduce any tremor in your arms. To paint details, hold the brush close to the tip, just behind the ferrule. This allows more control when working on small pieces or in tight spots. Leave your little finger free or gently placed on the model or the workbench to help stabilize your hand.

Most people believe that a brush is a simple tool. All you need to do is dip it in paint and slap it on, right? No. Depending on the task you want to perform, you'll use the bristles in different ways.

TIP: Perform precise work with the tip of the bristles. With a tiny amount of paint loaded on the tip, you can paint details, make freehand lines, and even create textures by stippling or striping. Although there is a wide belief that the finer the brush, the more accurate it is, this isn't entirely correct. The most important thing for a precision brush is to keep its tip pointed and maintained.

BODY: To cover a surface, we'll use all the hair of our brush. With the brush loaded with paint, make strokes in the same direction, until the target area is fully and thoroughly covered. With larger surfaces we must be patient because, due to thinning the paint, a single coat won't be enough. Let the paint dry and add successive coats until you get an even color. The size of the brush is also important. It will be difficult to use a No. 3 flat brush on small details, and you shouldn't use a No. 0 round on large areas.

EDGE: Use the edge of a brush for outlining. It works because only a tiny, stiff part of the bristles comes into contact with the edge of your model. This way you can manage straight, almost perfect strokes. Any size of brush can be used, as long as it is consistent with your work, but it's important to unload any excessive paint from the brush before using it.

Now that you know how to use each part of your paintbrush, you can get to work. A paintbrush is not a marker that perfectly covers an area with a single layer. Paint straight from bottle or jar often comes out too thick to use. Thin it to get a smooth consistency that flows well from the brush. While a milk-like consistency works well for airbrushing, you may want your paint a little thicker or thinner depending upon what you intend to do with it—fine lines, broad coverage, highlights, and do on. Once your paint is ready, dip the bristles in the paint, unloading any excess on a piece of paper towel. Brush strokes should be smooth and always in the same direction, following the contours of the model.

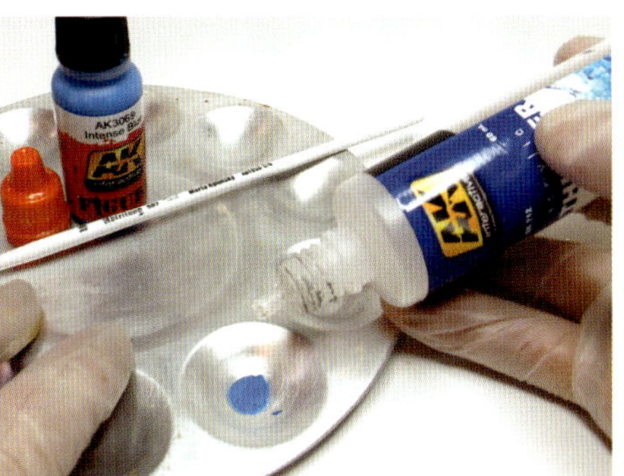

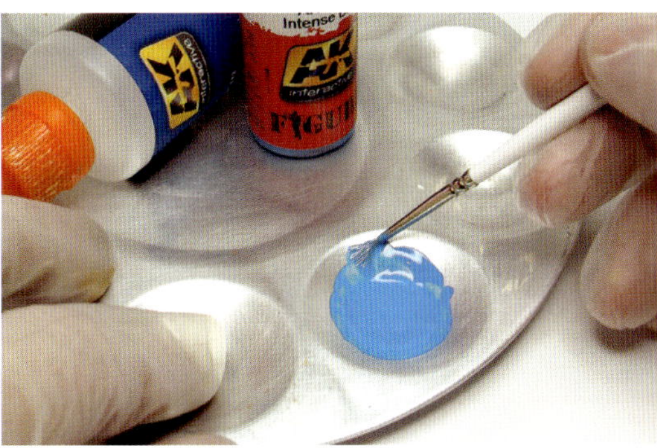

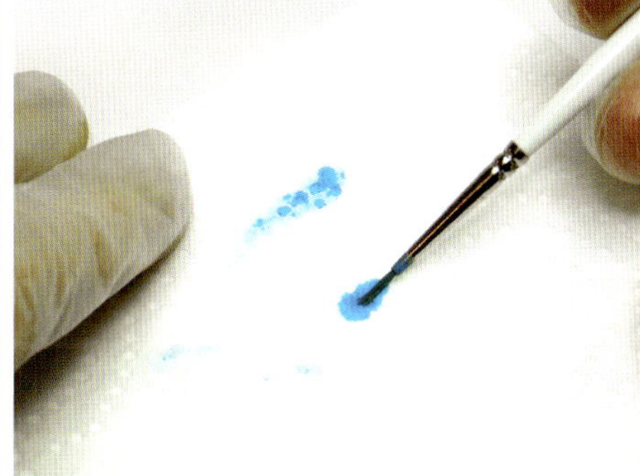

With acrylic paints, you'll notice that the spot where the paintbrush first contacts the model has less paint than where you lift the brush from the model at the end of the stroke. Thus, to achieve a perfect finish, the first and last point of contact must be altered with each brush stroke until the whole area is evenly covered.

Brushes require maintenance and care if you want them to last. When you finish a painting session, thoroughly clean your brush with soapy water, if you're using acrylics. Use the appropriate solvent if you're painting with enamels, lacquers, or artist oils. After cleaning the brush, dip it in preserver. This will considerably lengthen its life. Lastly, with your fingers, reshape the bristles into the proper shape while still damp. For instance, reform the fine tip of a round brush and make sure all the bristles of a flat brush are neat and uniform. If you have the protective tube, place it over the ferrule, but otherwise store all the brushes with the bristles up.

TIP

- Practice your strokes on a piece of paper or scrap from your model to develop your technique before applying it to your model.

Choosing an airbrush

Airbrushes paint surfaces by means of pressurized air. A compressor pushes air through a nozzle, drawing paint from a container, and turning it into a fine spray. Airbrushes offer a vast list of advantages, the most apparent being a smooth finish without brush strokes. You can create smooth gradients and more easily create paint shading effects that add depth and volume.

An airbrush can intimidate novice modelers. They're expensive and there's a lot of information you have to assimilate. But don't let that scare you off. There are many buying options, and while an investment, you don't need a top-end airbrush to begin with. Just one that is reliable with which you can achieve consistent results.

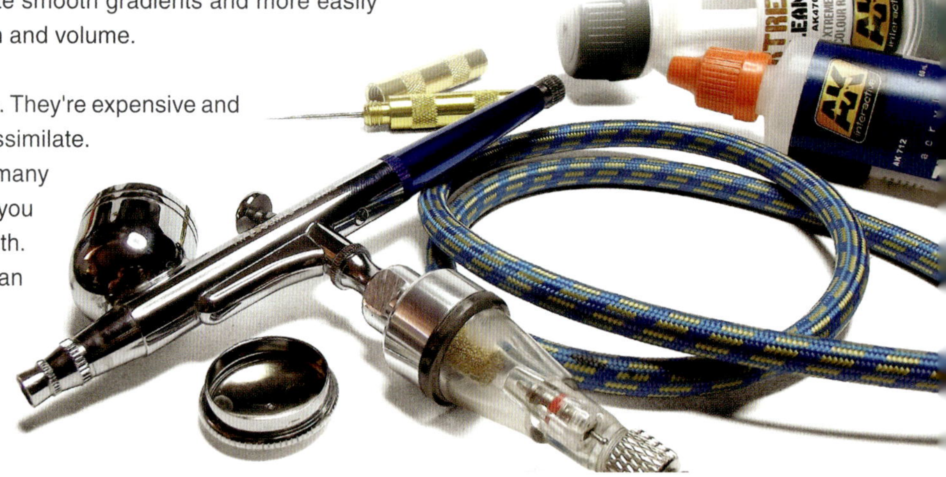

Seemingly complex, you can see that an airbrush can be broken down to a few component parts. The most important are listed below.

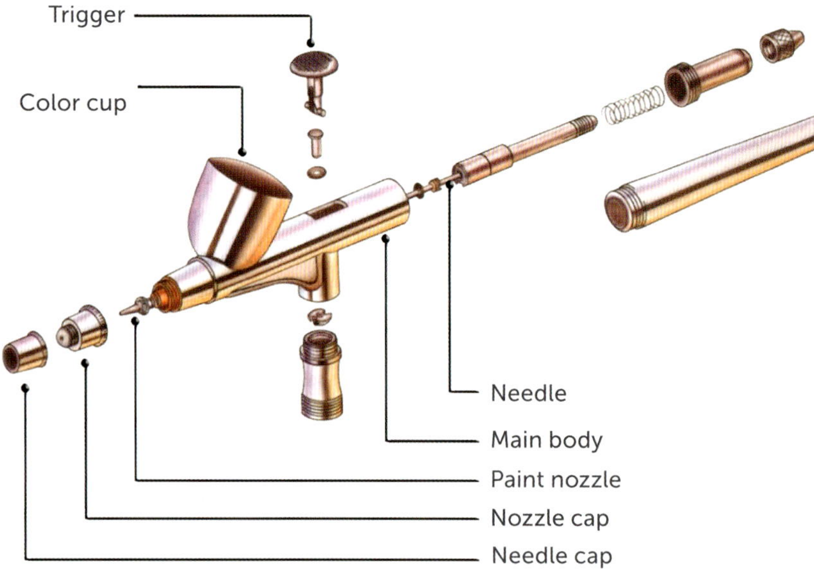

MAIN BODY: Everything else attaches to the main body. You hold the main body when painting with an airbrush. **TRIGGER:** The trigger controls the amount of paint moving through the airbrush. Pull all the way back for maximum flow; just a little for fine spray. **COLOR CUP:** Container for the paint. **NEEDLE:** It is the part that opens the nozzle for paint. Directly controlled by the trigger on a double-action airbrush. **PAINT NOZZLE:** The exit for the paint. The needle passes into it and regulates the amount of paint passing through. **NOZZLE CAP:** Covers the nozzle and is where the air and paint mix. It connects to the main body. **NEEDLE CAP:** Fits over the nozzle cap to protect the point of the needle.

When dismantling an airbrush, extreme care must be taken not to lose or damage the tiny and delicate parts it's made of. The most delicate part, and thus most prone to damage, is the "needle." Its pointed tip can be easily bent and distorted, causing the airbrush to malfunction and produce a poor finish. The needle must always be kept perfectly clean and lubricated per the owner's manual as the overall performance of the airbrush strongly depends on it.

There are many cleaning tools available, ranging from micro-brushes to ultrasonic cleaners, complemented by purpose-made fluids and mediums.

Airbrushes come in a variety of styles including siphon feed and gravity feed varieties. Some have a pen-like body style, like the ones seen below, while others are fitted with a pistol grip. Lastly, airbrush mechanisms can be categorized as either single- or double-action. All of these differences change how paint is applied, how easy an airbrush is to operate, and how difficult it is to clean.

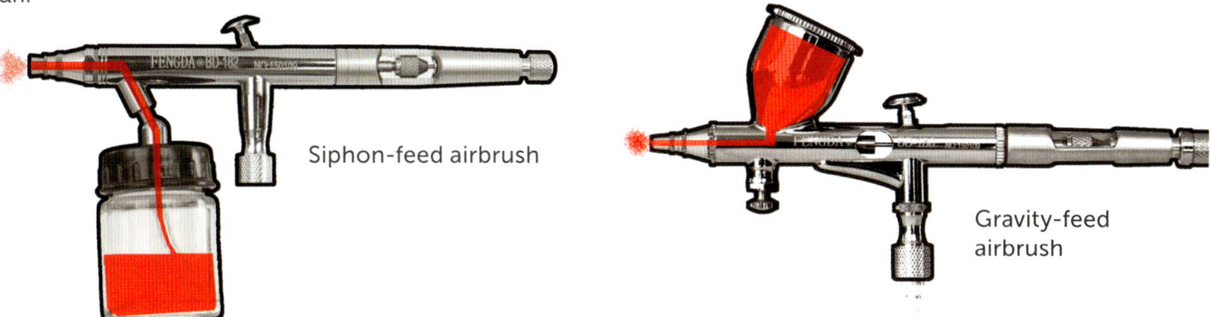

SIPHON FEED: Siphon-feed airbrushes have a removable cup suspended under the main body from which the paint is siphoned when the trigger is pressed. Great for working on large surfaces because they can be loaded with more paint. Easy to clean. The position of the paint bottle limits your ability to get close to a surface, and makes it less useful for precision painting.

GRAVITY FEED: Gravity-feed airbrushes usually have a color-cup on top of the main body from which the color is pulled into the needle chamber by gravity. More difficult to clean than a siphon-feed airbrush. Ideal for precision painting since the position of their color-cup allows you to get closer to the model.

SINGLE ACTION: This airbrush type operates a lot like a spray-paint can. The trigger in a single-action airbrush only controls the airflow. When the trigger is pushed down, air passes through the airbrush and pulls the paint out of the nozzle. When the trigger is released, the airflow stops. You preset the paint flow by adjusting either the nozzle or the needle.

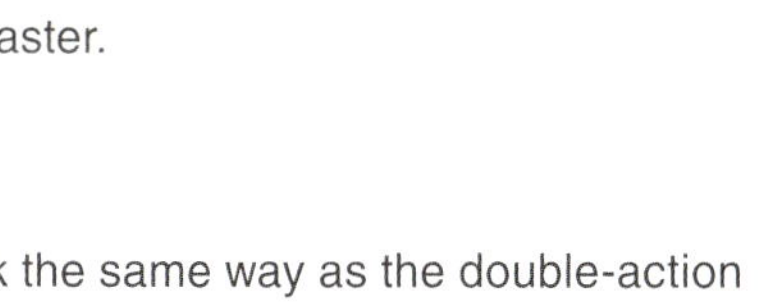

DOUBLE ACTION: On double-action airbrushes, you control airflow by pushing the trigger down; you regulate the paint flow by pulling the trigger back and moving the needle out of the nozzle. Double-action airbrushes offer more flexibility for creative results depending upon airflow and paint consistency—slightly stippled to a glaringly smooth gloss finishes aren't out of the question. But double-action brushes take time and practice to master.

TURBINE ACTION: These rare airbrushes work the same way as the double-action airbrushes. However, they're uncommon among scale modelers and are used mostly by experienced professionals.

After you buy a new airbrush, you have to tune it and practice with it. A good idea is to buy a sketch pad of heavy paper and practice on it before moving to your model.

TIPS

- Ask an experienced modeler for help choosing an airbrush.
- Always have thinner, water, and a good airbrush cleaner nearby.
- Clean your airbrush between colors and especially after each session.
- Your airbrush has delicate parts. Don't bump or drop it, and store it in a safe place.

Airbrush compressors

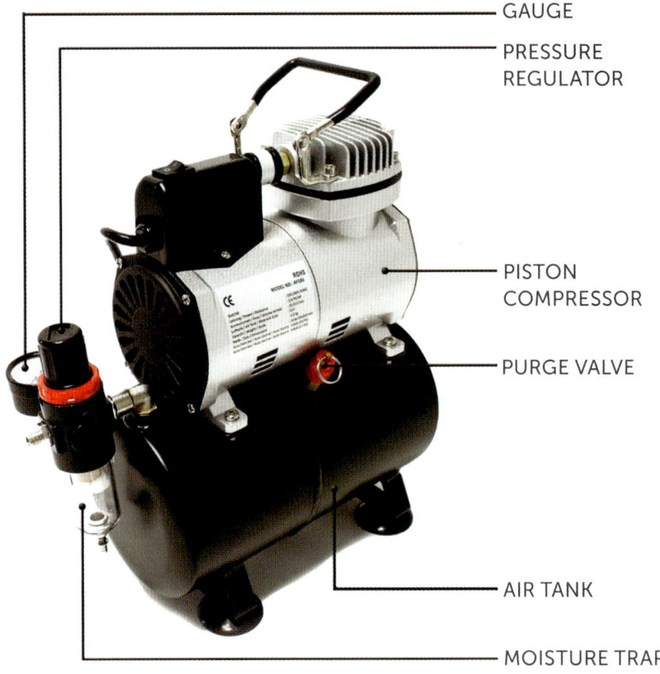

An airbrush needs compressed gas to operate. While some modelers like to use compressed carbon dioxide or nitrogen, you have to recharge the propellant tanks, and you may run out of pressure in the middle of your project.

The better option is an electric air compressor. A compressor with a large tank will run for a while, fill that tank, and then shut off. It only kicks on intermittently as you use the air; it may not kick on at all if the tank is big enough. However, those size compressors tend to be expensive. A more affordable choice is a small hobby compressor. Often you can find a compressor designed to work with the airbrush you purchase. Or you might buy them as a set.

Whatever compressor you decide to buy, make sure it comes with a pressure gauge, pressure regulator, and moisture trap. You want these to control the air pressure supplied to your airbrush and to keep unwanted condensation or oils from spitting onto your model.

Commonly, modelers opt for piston or diaphragm compressors. Either works well, though they do make noise. Silent compressors have become more popular recently.

PISTON AIR COMPRESSOR
These compressors are the most common for scale modeling. They are lightweight and easy to transport and maintain. Piston air compressors offer many efficient features, making them popular with new modelers.

DIAPHRAGM AIR COMPRESSOR
Cheapest of all the compressor types, diaphragm compressors provide continuous air flow, which makes it difficult to work with on fine details. On the other hand, the air does not come in contact with the pressure mechanism and remains oil residue-free.

SILENT COMPRESSOR
Compact, easy to use, and safe for you and your model. Silent compressors work well for scale modeling. They're quiet and ideal for indoor use. They're usually quite affordable, but you can find more expensive models.

COMPRESSED GAS
Compressed air, carbon dioxide, or nitrogen can be found in cans and pressurized tanks. These propellants can be purchased and refilled, and provide steady pressure for your airbrush. However, the gases can be toxic, the cans must be kept away from high heat, and you need a good-quality valve for accurate pressure control.

When you start painting with an airbrush, the first challenge you have to conquer is air pressure. Excessive pressure will cause the paint to reach our model with great force, making it difficult to control. High pressure will result in overspray or splashes (1).

On the other hand, low pressure will spit small droplets of semi-dried paint onto your model, giving unwanted texture to your finish (2). If the pressure seems right and the spitting effect persists, check if your airbrush is clogged.

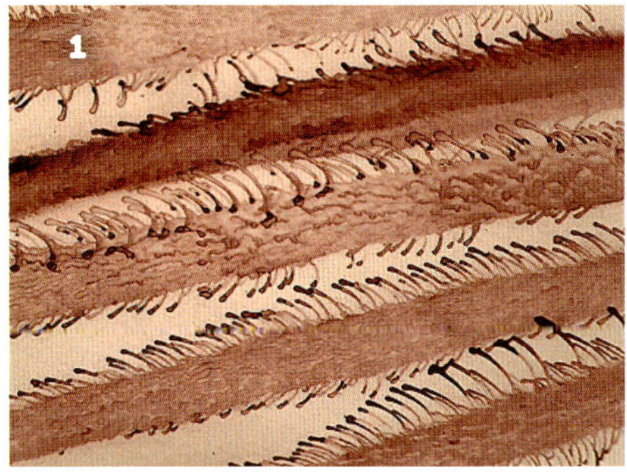

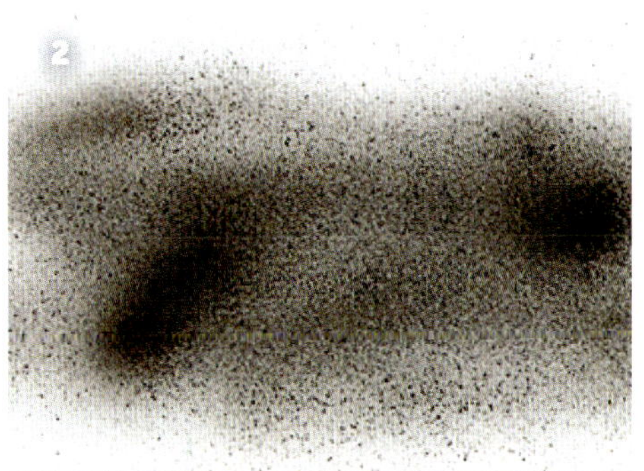

To a novice modeler, it may seem overwhelming to take into account so many variables. Be patient and practice. Become familiar with your airbrush before using it on a model. Remember that the ideal pressure for airbrushing is 14 to 22 PSI. With a clean airbrush and properly thinned paint, you should get good results every time.

TIPS

- If you find yourself doing long painting sessions, consider investing in an air compressor with a large tank that will kick on fewer times during a session.

- When first starting to airbrush, invest in a good compressor rather than a good airbrush. As your skills improve, you can buy a better airbrush. If wisely chosen and maintained, a good compressor will last for many years.

Using an airbrush

As you know, there are basically two types of airbrushes: single-action and double-action. Now, let's see how each type is used. All processes described concern both types, the only crucial difference between the two being the trigger. First, connect the airbrush to the compressor. Pour your thinned paint into the color cup or bottle, and you're ready to paint. Using the regulating valve, adjust the pressure on your air compressor as desired.

Here's where you meet the differences between the two types of airbrush.

The trigger on a single-action airbrush only pushes down. When pressed, it allows the air to flow through the body, meet the paint, and exit the nozzle in the form of a spray. Throughout this process, the only control the user has is on the flow of the air—it's either flowing or it's not. Control comes from adjusting the nozzle and needle, or pressure at the compressor.

On a double-action airbrush, the trigger moves in two directions: down and back. Pressing the trigger down controls the airflow. Pulling the trigger back controls the flow of paint.

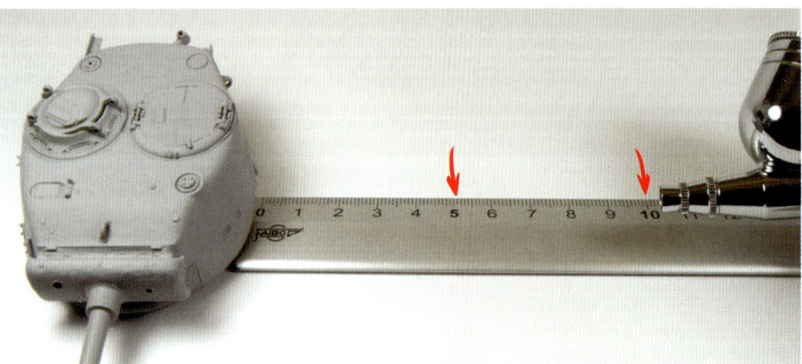

Hold the airbrush about 5 inches (or 10cm, as shown) from the model. You can adjust closer if necessary, but this a good place to start.

Always begin spraying off the model. Press (and pull, for double action) the trigger, aiming near the model, then gently move the airbrush across the model's surface. Remember, always begin and finish spraying off the model.

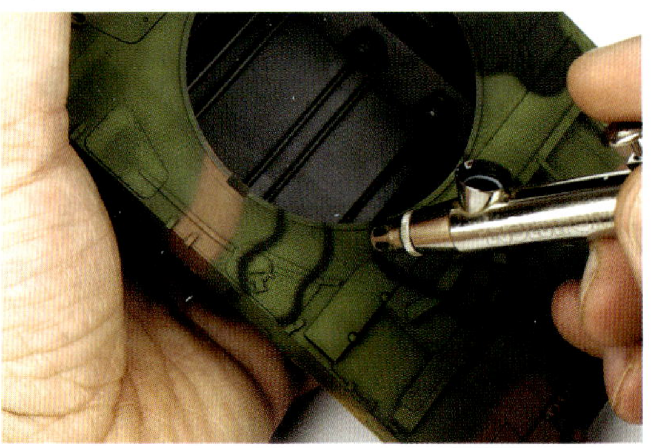

Cover the surface with successive thin layers, leaving each area some time to cure before making the next pass. Use either a zigzag or linear pattern, moving both horizontally and vertically, over the surface.

When using a gravity-fed airbrush, paint should be mixed and diluted in a bottle or container before it's transferred to the color cup.

As your experience with the airbrush grows, you'll be able to develop your own personal style and standardize your preferences.

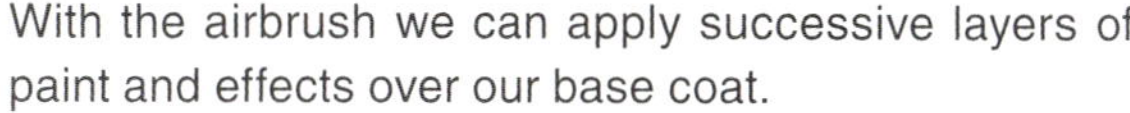

With the airbrush we can apply successive layers of paint and effects over our base coat.

Paint schemes often have one of two finishes: hard-edge, on the left, or soft-edge, on the right. To paint a hard-edge scheme, mask the base coat with tape. Airbrush the new color, spraying perpendicular to the surface. Remember to avoid causing runs or pooling.

To apply a soft-edge scheme, mask with masking putty, or poster putty such as Blu-Tac. These masks allow a feathered outline. Alternatively, you can use your airbrush freehand, with no masking at all. Airbrushing freehand can be a freeing experience, but it's also difficult to get the results you want without practice.

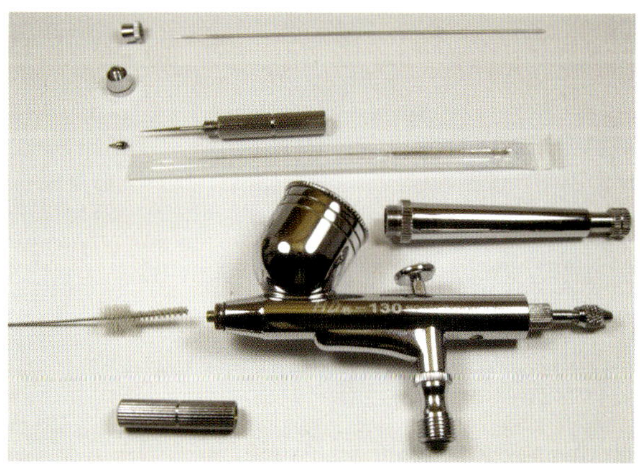

You can use the airbrush with the nozzle-cap on or off. With the cap off, narrow and even ultra-fine lines can be achieved, which is ideal for precision painting, such as complicated camouflage schemes.

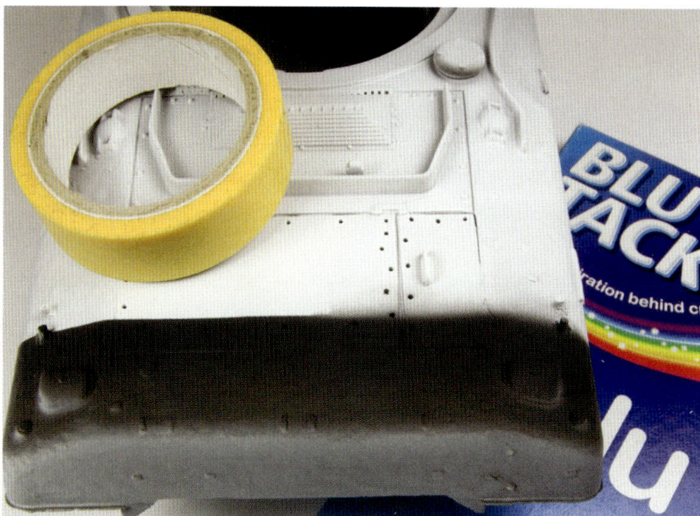

One of the most common problems one can face painting with an airbrush is the formation of **orange peel** on the model. This happens when part of the color dries in contact with the air before it reaches the model. The usual suspects are low air pressure, a clog in the airbrush, or a very dry or hot environment in your workshop. If it's the last, you may want to invest in a humidifier, and try to keep the heat at room temperature (68° F).

Clean your airbrush between color changes and at the end of each painting session. Not only does it keep your airbrush working well, but it also extends its life span. Follow these simple rules: Disassemble your airbrush following the manufacturer's instructions. Thoroughly clean the nozzle, nozzle cap, and needle (top) with a paper towel wet with acetone or other strong cleaner. The main body (center) can be carefully cleaned with a brush. The area between the body and the mouthpiece (photo bottom) can be treated with a brush or a toothbrush.

TIPS

- Practice on a scrap piece of paper or other material before moving to the model.
- Proper thinning, adequate air pressure, and a clean airbrush are the keys to airbrushing success.
- There are ultrasound devices at very affordable prices that help to the clean inside of the airbrush. It's a useful tool that will come in handy.
- If you're using paints that require aggressive solvents for thinning and cleaning, consider investing in a cheaper secondary airbrush for use with them.

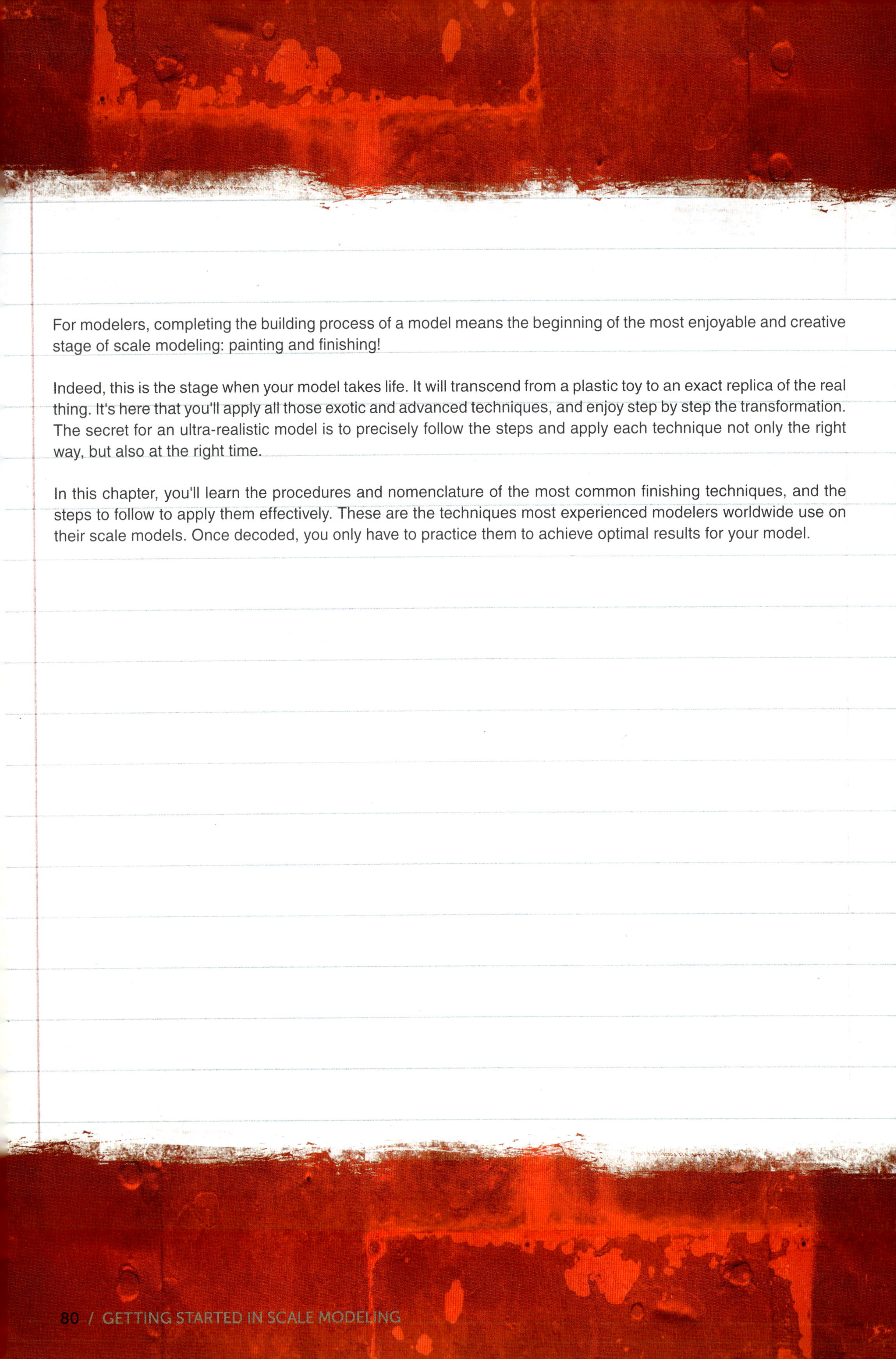

For modelers, completing the building process of a model means the beginning of the most enjoyable and creative stage of scale modeling: painting and finishing!

Indeed, this is the stage when your model takes life. It will transcend from a plastic toy to an exact replica of the real thing. It's here that you'll apply all those exotic and advanced techniques, and enjoy step by step the transformation. The secret for an ultra-realistic model is to precisely follow the steps and apply each technique not only the right way, but also at the right time.

In this chapter, you'll learn the procedures and nomenclature of the most common finishing techniques, and the steps to follow to apply them effectively. These are the techniques most experienced modelers worldwide use on their scale models. Once decoded, you only have to practice them to achieve optimal results for your model.

4. Painting techniques.

Where to start

During assembly, you must plan how you'll approach the painting process. Study the instructions (1).

While some kits can be assembled completely before being painted, others will require painting parts separately or as a subassembly. Then they'll be added to the final model (2).

Prepare the model surfaces prior to painting. After the manufacturing process, model parts tend to have oily residues that prevent the paint from holding well. Clean the parts with soap and water (3).

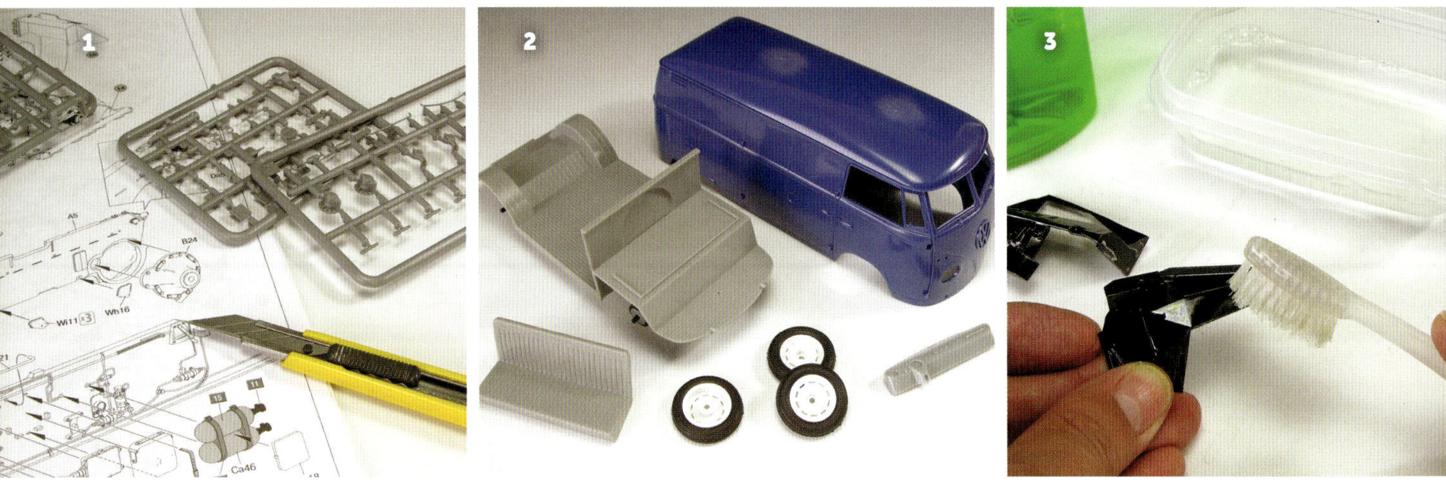

Studying the instructions and deciding which pieces to assemble or paint separately is crucial for a good result.

Many models include parts made from various materials. In addition to a good cleaning, prime all the parts before painting.

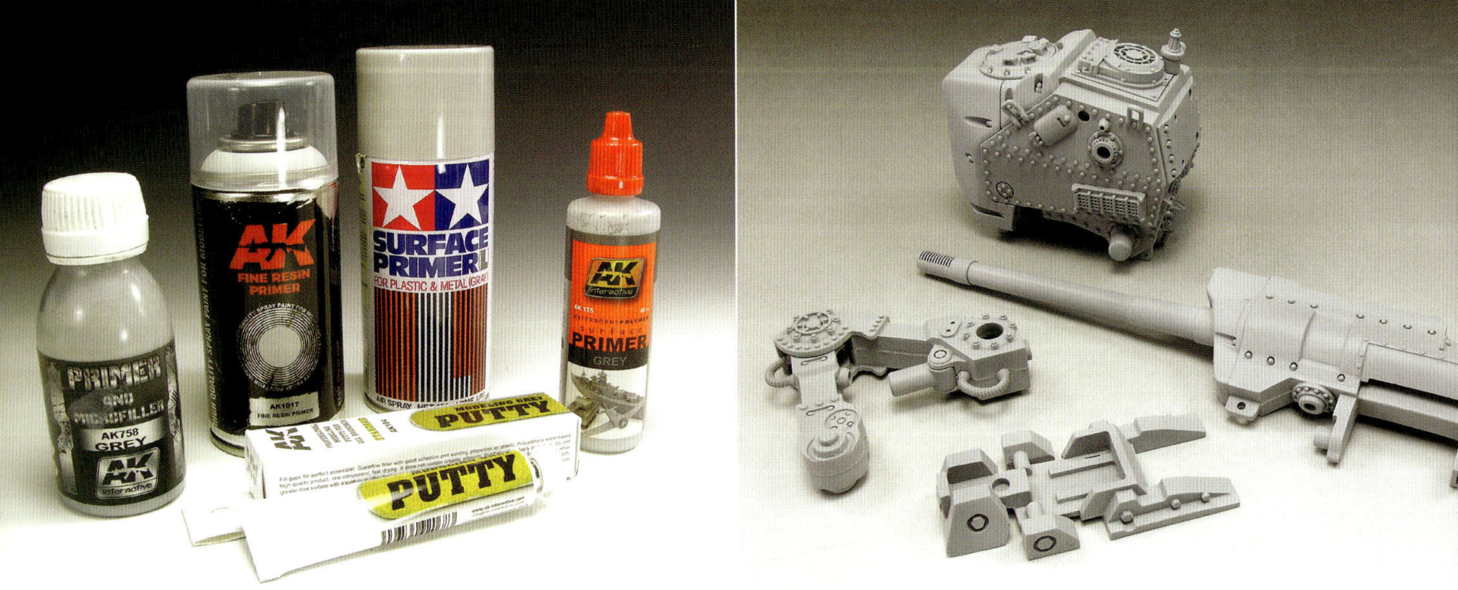

Painting a model, no matter the subject, follows a certain logic. For example, painting wheels or accessories separate from the rest of the model is a logical way to work. Also, leave the fragile parts, such as rearview mirrors and antennas, for the end.

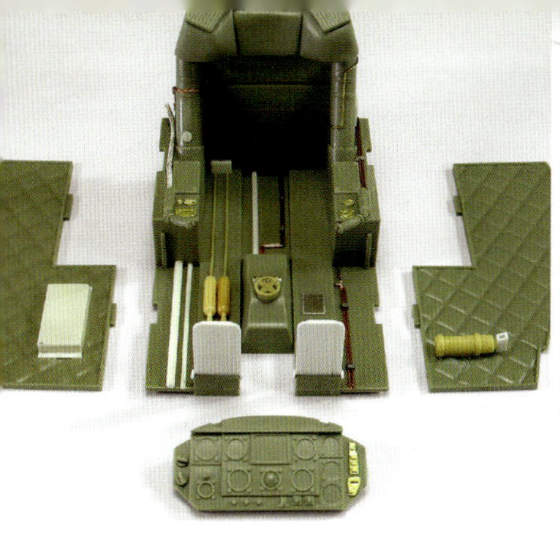

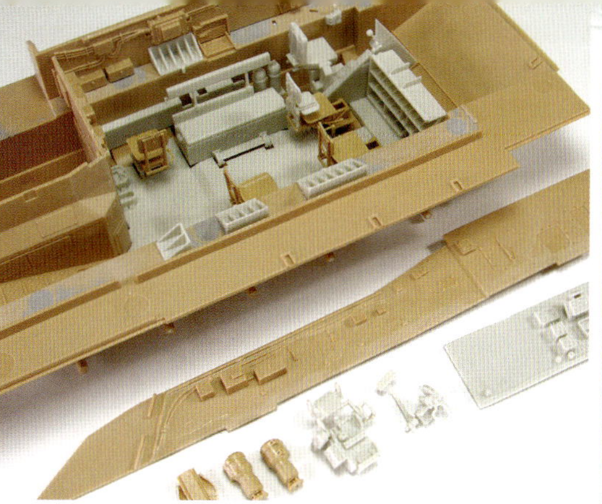

Obviously, for models with interiors, it makes sense to finish the interiors before assembling all the parts. In these cases, carefully protect the interior with masking so it isn't damaged during the painting and finishing to come.

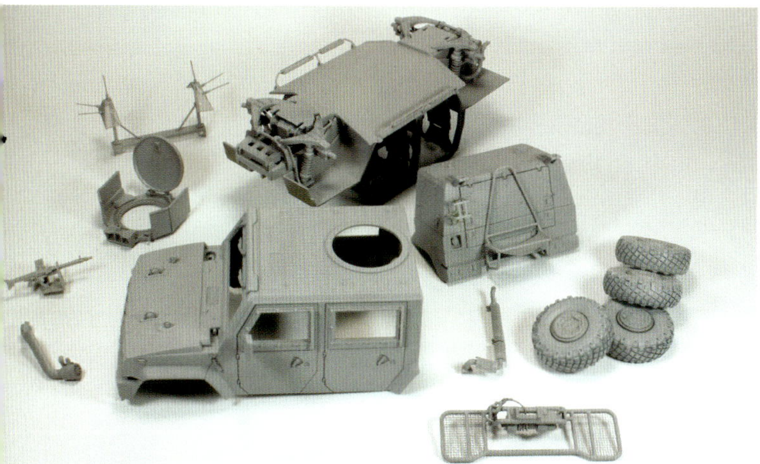

When the main subassemblies are complete, you'll have a number of smaller pieces, such as hatches, tools, crates, and weapons, you'll have to paint separately. It's fundamental to plan how to hold them in order to paint them completely.

Parts that have a solid base that will be hidden once it's added to the model can be attached to a surface with double-sided adhesive tape (4). Attaching a part to the end of a short rod with some Blu-Tac allows you to maneuver the part for better coverage (5). For other parts, it may make sense to drill a small hole and insert the tip of a toothpick. If you'd rather not drill the part, cut the end of the toothpick flat and attach it with a drop of super glue (6).

For the larger parts (7) and subassemblies (8), use your hands while wearing latex gloves. Make sure to paint from all angles to get even coverage.

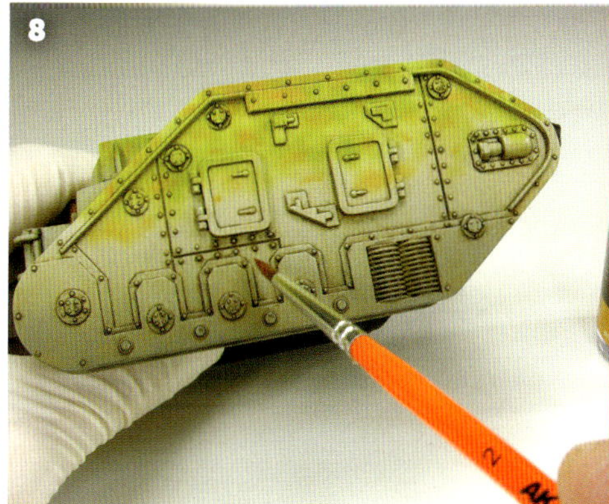

Primer

The first step before starting with the painting process is to apply a primer. Although it seems insignificant, priming your model is important for the following reasons:

- Helps paint adhere to the surface of the model
- Unifies the pieces and materials of your project. This is imperative in the case of multimedia kits, with plastic, resin, and photo-etched-metal parts
- Provides a homogenous tone for base color
- Makes imperfections more noticeable, letting you fix them before you start to paint

There are many types of primers available, from spray cans to primers you can hand-brush and airbrush. And they come in a number of colors. Consider what type is best for the model you're building.

SPRAY-CAN PRIMER

Best for models that do not incorporate a lot of detail, or larger models and surfaces. Spray cans offer speed since you don't need to clean a tool or add more paint. However, spray cans lack accuracy and offer limited control over the process. Sometimes, depending on the quality of the product, the finish might not be as smooth as desired. Make sure the parts to be primed are clean and free of grease or other residues. Shake the can for the time indicated on the can to mix the primer. Spray from a distance of 6 inches to 12 inches in short, rapid applications. Remember, begin and end spraying off the model.

AIRBRUSH AND PAINTBRUSH PRIMER

Priming with an airbrush gives the best results, hands down. You can apply this primer with a paintbrush, but the finish probably won't be as smooth and even as it is with an airbrush.

Follow two fundamental steps when priming with an airbrush:

1. Use a clean airbrush and set the air pressure to 12 psi. Thin the primer 1:1 to avoid clogging. Transfer the primer from its container to the color-cup with a brush to avoid bubbles.

2. Evenly spray the primer over the model, altering directions, making sure to completely cover the model. Allow a coat to thoroughly dry before applying another. It's always preferable to apply multiple fine layers rather than a single thick coat.

Another factor to take into account when choosing a primer is its color. The most common primer colors are:

BLACK: Provides a more muted finish. Works well for figures or models with diverse surface levels and crevices such as robots, armor, and thick fur
WHITE: Brightens finish, enhances colors, and transfers luminosity. Great for clothes, light skin tones, and smooth surfaces
GRAY: The original primer color. Provides a neutral base
COLORS: Colored often to represent a primer color used on a real vehicle. They are widely used for historic vehicles and aircraft, where color accuracy is essential and the primer will be visible due to paint scrapes, damage, or other weathering.

FRAMING WITH PRIMER

A popular figure-painting technique is called **framing**. Framing simulates the lighting a figure will have. For example, you can have a figure painted as if the light-source is coming from the right, thus all the features on the figure's left side will be painted in darker colors. Framing can be more even more dramatic if based on a similarly primed figure. Follow the steps below to frame a figure with primer:

1. Thoroughly clean the figure with a toothbrush and alcohol. Let it dry thoroughly.

2. Apply a uniform coat of black primer Take care that everything is completely covered. Again, use a 1:1 thinner-to-primer ratio.

3. To frame the figure, airbrush a thin layer of gray or white primer, spraying from the direction of the imagined light source. In this example, the figure has been framed with light from overhead. You can see where the shadows will be deepest. Use three parts thinner for every one part primer for this step.

TIPS

- Don't forget to clear the spray-can nozzle by turning it upside down and pressing the valve until only propellant escapes.
- After you've primed your model, touch it only when wearing gloves to avoid transferring dirt and oils or marring its surface.

Base color, shadows, and highlights

Trying to learn about painting techniques for scale modeling, one will meet concepts that may seem complicated and difficult to comprehend.

Start by painting a scale model a base color. This will be the dominant color, but also the base for all subsequent shades and tones necessary to achieve the final look. Although all this may seem obvious, it's important to have these terms clear in your mind, as they'll constantly appear in explanations and examples in scale modeling publications. To choose the right base color, we must investigate our project and choose one closest to the documentation, photos, or other reference sources.

Here, we're painting an ambulance from World War II, and we've chosen, based on the related documentation, the correct base color, olive drab.

Usually, you'll airbrush the base color on the entire surface of the model, but on occasion you might need to paint only a part. In any case, always use the same methodology for the base color. Either painting with a paintbrush or an airbrush, we gradually cover the surface of the model to create a solid layer of color, taking care not to leave any areas or spots unpainted, as this may later affect the overall look of our model.

So far it's clear that a base color is the first, solid layer of color applied on a model. This layer can also be enriched with some lighting effects. The two main elements that will accompany us throughout our journey in painting scale models, are the **highlights** and the **shadows**. The highlights are the tones or colors that we use to simulate areas that are directly in light or more affected by the light. Shadows are the colors we use to simulate areas less affected by the light.

In general, when painting with a brush, you tend to first apply the highlights and then add the shadows. When using an airbrush, first apply the shadows and then the highlights. As we saw in the chapter for color theory, to get a tone, it's not enough to randomly mix colors, nor just add white to lighten and black to darken.

To effectively create the shadows, we can gradually add a darker shade to the base color, and also a small quantity of its complementary color, to improve realism.

We must always follow this rule, although sometimes the complementary color that results may seem peculiar.

In our example, we add a small amount of black and violet to get our new shadow color.

When creating the highlights, we can mix the base color with its analog color, plus a little bit of white and a small amount of the complementary color to enrich the mixture. We can also look for a similar color, but one that already contains white in its composition. According to the analogs chart, our model will look cold or warm, depending on our choice. For the ambulance, a small quantity of ocher was added to the base, lightening the color.

How much of the base color shows on the final model can vary greatly depending upon how many other colors are layered on top of it, such as with this complex digital camouflage.

The base color provides the general feel of your model, but without any volume or depth to the colors. The highlights and the shadows are elements that provide a more dimensional effect, which in turn, lends realism to your model. Making good use of these effects, your projects will come to life. However, combining what you've learned with other weathering and finishing techniques, you'll make your model stand out. In scale modeling, the lighting effects work the same way as in an illustration or in a 3-D design: They enrich and provide interest to the all areas of a model.

TIP

- The color wheel is a reference you should have at hand when mixing colors.

Lighting overview

In a previous chapter, you learned about lighting your workbench and the importance of working in good lighting conditions. Now, you'll explore the different types of light—not those over your workbench, but the effects applied to your model. At the same time, you'll see different techniques to effectively re-create the effect of light your model. In the past, modelers didn't consider adding lighting effects to their models. Today, modelers take a more artistic approach to scale modeling, making lighting effects a must for every build. Highlights and shadows are two of the basic factors to achieve a realistic result. To what degree you use the techniques is always a matter of personal taste. One can refer to the famous painters of the Renaissance, to see how personal taste and style affect the way they rendered light.

Light can be one of the most important and more complex stages to complete while painting a scale model. The color of an object is variable—it changes according to the environmental condition to which it's exposed. The most important factor affecting a color is ambient light. The facade of a house will look very different under the noonday sun than it does at sunset.

A light source can dramatically change the tone of an object depending on its intensity or the color itself.

Another crucial factor to take into account when considering light effects is scale. Study how light illuminates full-size vehicles and buildings. Find photo references of the full-size version of the model you're building to see how light plays off its surface in different settings. Then use your knowledge of colors to simulate that effect at scale. Too much, and the model will look like a cartoon. Too little, and it will appear flat, like a toy. And remember, you won't get it completely right the first time. But that shouldn't keep you from trying. The next model you build will be better. And the next, even better.

Now, lets' see some more about the two types of lighting that can be represented on your model or figure.

ZENITHAL LIGHT

This the most common lighting you find on models. Zenithal light is generally located above the model and perhaps a bit to the right or left for some volume. The light loses intensity as it descends along our model. You can use zenithal light to attract the eye to a particular detail or part of the model. The more illuminated and detailed an area is, the more it will stand out. The darker and more simple, the more it recedes into the background.

FOCAL LIGHT

This is a much more artistic representation of light, since we are simulating ambient illumination from a different source, usually artificial, whose direction and intensity you choose. The light of a fire or a light bulb are sources that can illuminate your model. Focal light, in general, must be very close to the subject and intense, casting light in a confined area and at the same time creating stark shadows and generating a lot of contrast. Focal light, apart from its varying angle and intensity, can also have color! The blue light emitted by a lightsaber, for example. Representing focal light is a difficult skill, but if mastered, can create stunning models.

Sometimes you'll want a model to be represented with natural light (zenithal), and at the same time, illuminated by an artificial light source (focal). In such cases, you have to undertake the difficult task of combining both types of light and make the result look credible. Take into account the zone(s) of brightest light and the zone(s) of maximum shadow. This contrast, simple as it is, adds coherence to your work. The zone of brightest light is the part of the model most exposed to light. Therefore, it's the area that will naturally draw the viewer's eye. In contrast, the zone of maximum shadow sees little or no light, and often goes unnoticed and usually has less detail.

TIPS

- Consider your lighting choices carefully; the realism of your models depends on them.

- Light should come from a consistent source. If a project consists of several models, use the same light source for them all.

Gradients

When modelers dive into contemporary painting and finishing techniques, they usually come upon unfamiliar terms that, if misunderstood, can lead to confusion and frustration. One such term is **gradient**.

In plain words, a gradient is a directional change in the intensity of a color or a directional change from one color to another. For example, a gradient change from yellow to red would display a transition from yellow through orange to red.

COLOR GRADIENT

A color gradient makes a smooth transition between two colors anywhere on the color wheel. This kind of gradient is used mainly in fantasy figures, in which the imagination leaves room for gradients that are rarely found in nature, such as red to green.

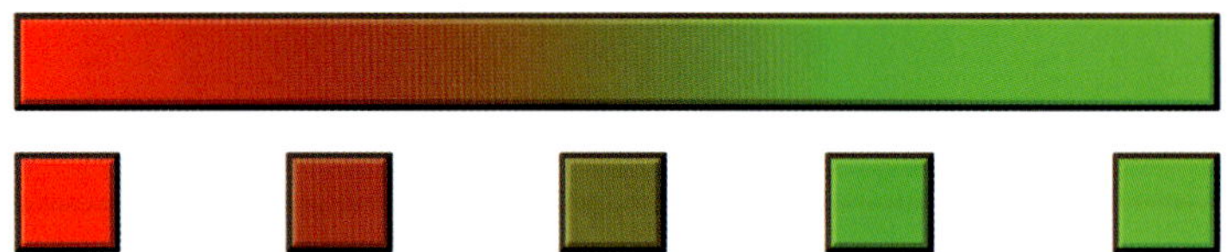

INTENSITY GRADIENT

Also called a light transition, this gradient smoothly translates a color from its lightest shade to its darkest. You find this type of gradient on all types of models and represents the effect light has on a color. This is the type of gradient you'll focus on and use most.

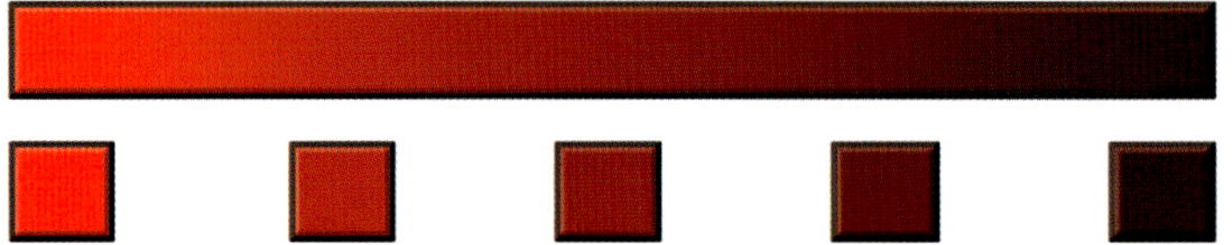

LIGHT TRANSITION

When facing the task of painting a light transition for the first time, many modelers end up adding white to reach the lighter shades and black to reach the darkest. As we know from color theory, that's not a good idea. Adding white creates an unreal, faded tone, while black will make the color look dirty.

To achieve an effective gradient, we have to use the intermediate of our base color, according to the color wheel. To make things easier, you can buy a ready-to-use gradient paint set. There are sets available for many different colors. For our example, let's use a red gradient set. Use an airbrush for this kind of painting because it will allow you to make perfect transitions.

1. We're going to paint a typical shield (scutum) of a soldier from ancient Rome. They were painted vivid red. First, clean the part and apply gray primer for a neutral starting point.

2. Airbrush the base color onto the shield. Again, thin coats are best, gradually building up to a solid, even color overall.

3. Now, using the next-to-darkest color in the set, airbrush the first shadow. Spray a light coat from below to get a subtle transition from the base color midway up the shield to the darker color. Then move on to the darkest color. Again, spraying from below, cover the bottom of the shield and work about half way up the previous color, making a smooth gradient.

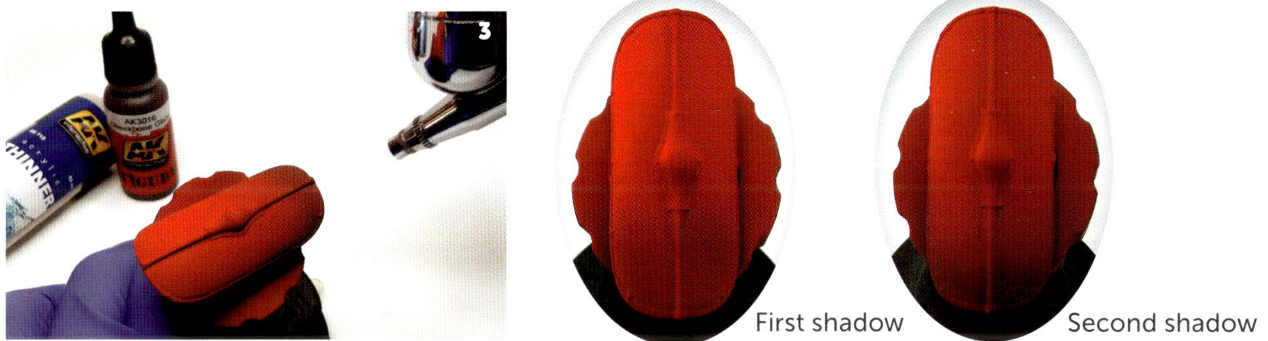

4. For the top of the shield, follow the same steps with the lighter shades, starting with a tone just a bit lighter than the base color. Finish with the lightest color. Spray from the top of the shield, downward. The first light tone should cover less than the upper half of the shield, and the second tone even less than the previous one.

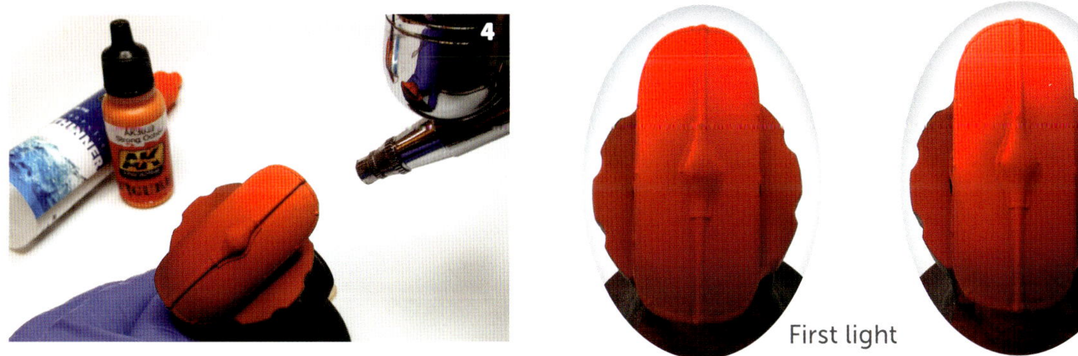

The work is finished with a pinwash outlining both the light and shadow to delineate the transitions and add realism. A gradient can profoundly improve a model. Here, the flat, unremarkable shield we had at the beginning now pops.

TIP

- Use a 3:1 thinner-to-paint ratio to control the gradient transition. Better to apply multiple thin layers with the airbrush for a smooth gradient rather than end up with defined lines.

Color modulation

With color modulation, you exaggerate the effect of light so your model looks more like the real thing. Assume you have a fixed light source coming from above and slightly off to one side—where exactly is up to your personal taste—to provide a more dimensional appearance.

You don't only modulate when you apply highlights and shadows, but when you highlight an area already finished with glaze and filters. Although it seems complicated, color modulation is easier to perform than it sounds.

First, you prime your model neutral gray. You'll use your airbrush with air pressure set to about 12 psi. Thin your paint until it has a milk-like consistency, and add a retarding medium if you're using acrylics (10 parts paint to 1 part retarder). This will prolong the paint's curing time, leaving you more time to work on the model.

You can make our own color mixes using what you've already learned, or you can purchase a ready-to-use modulation set. Let's use a modulation set for our example below.

BASE COLOR

Airbrush multiple thin coats, allowing enough time for each coat to dry. Make sure to cover the entire surface of your model. When finished with this step, the base color of our model must have a solid, even look.

SHADOWS

The level of shadow you apply depends on your personal taste. Airbrush a single dark tone for the shadow or make gradient transitions until the darkest tone is reached—you decide. Whatever your choice, make sure to cover the model's lower areas and anywhere that wouldn't get much light.

HIGHLIGHTS - LEVEL 1

In this step, you apply the first highlights on your model. For this example, just two lighter tones than the base color are used. The number of tones you'll use on your model can vary according to scale—the larger the model, the more tones you can use on it.

Focus on covering areas with the greatest exposure to light, keeping toward the center of panels. The second tone should blend with and not cover the first. The same goes for all subsequent tones.

HIGHLIGHTS - LEVEL 2

To further enhance the lighting effect and achieve a more realistic appearance, you can apply an extreme highlight tone to the corners and edges of the model. Focus your attention on those areas only. The number of tones you will use on this step also depends on the scale of your model, and the level of detail you want.

The brightest highlights and darkest shadows were added with a paintbrush. This gives you more control of paint placement on very specific areas. Don't be afraid to switch back and forth between tools. You'll begin to understand that all your modeling tools can be used in conjunction with each other, and you'll recognize when to use one tool and technique and when not to.

This tank has gotten a pretty extreme makeover. It's gone from a flat, lifeless model to something that looks more likely to drive away. By imitating the behavior of light, exaggerating it to diminish the scale effect, you'll create a sense of motion and immediacy that your model would otherwise lack. This technique can be used effectively on any model in any scale, be it a plane, armored fighting vehicle, car, or figure.

TIPS

- Use a hair dryer to shorten drying times. Be careful! Excessive heat can warp plastic and resin parts.
- Color modulation sets make life easier, but making your own modulation tones can offer greater color variety to work with.

Panel lightening

Panel lightening is a technique originating from illustration and helps make your model look less flat, adding dimension to individual areas.

An advanced technique, you can lighten panels to complement other techniques. While an airbrush is the most efficient tool for the job, you can also use a paintbrush and artist oils thanks to their superior blending properties. Because this technique doesn't faithfully simulate reality, it allows you to take artistic license and vary the tone of the panels substantially. Nevertheless, it's widely used for its eye-catching results.

As the name implies, vehicles or surfaces with many panels are perfect for this technique. In its simplest form, panel-lightening is about saturating the center of each of the panels, as seen on this ambulance, gradually lightening the base color from the center of the panel to the edges. The effect is a faded color that enhances the overall look of the model. At the same time, you're preparing the surface for more weathering and finishing.

Panel lightening can be also performed with a paintbrush by dry-brushing the center of each panel with the desired color.

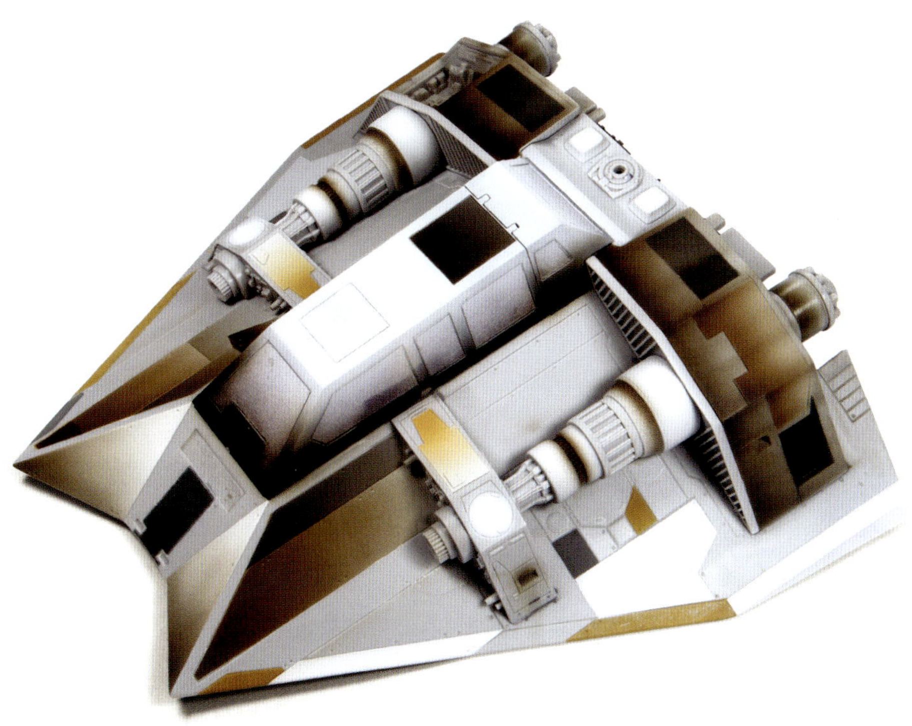

Panel lightening is perfect for vehicles with large, smooth, and well-defined panels. When creating lighting effects once you've lightened panels, make sure to use different colors than you used for the panel lightening and always remember where the position of your light source.

The effect produced by this panel lightening can be striking. Initially, the transitions may look sharp, but once softened with different weathering effects or even with a transparent layer of paint, the finish will look more feathered and natural.

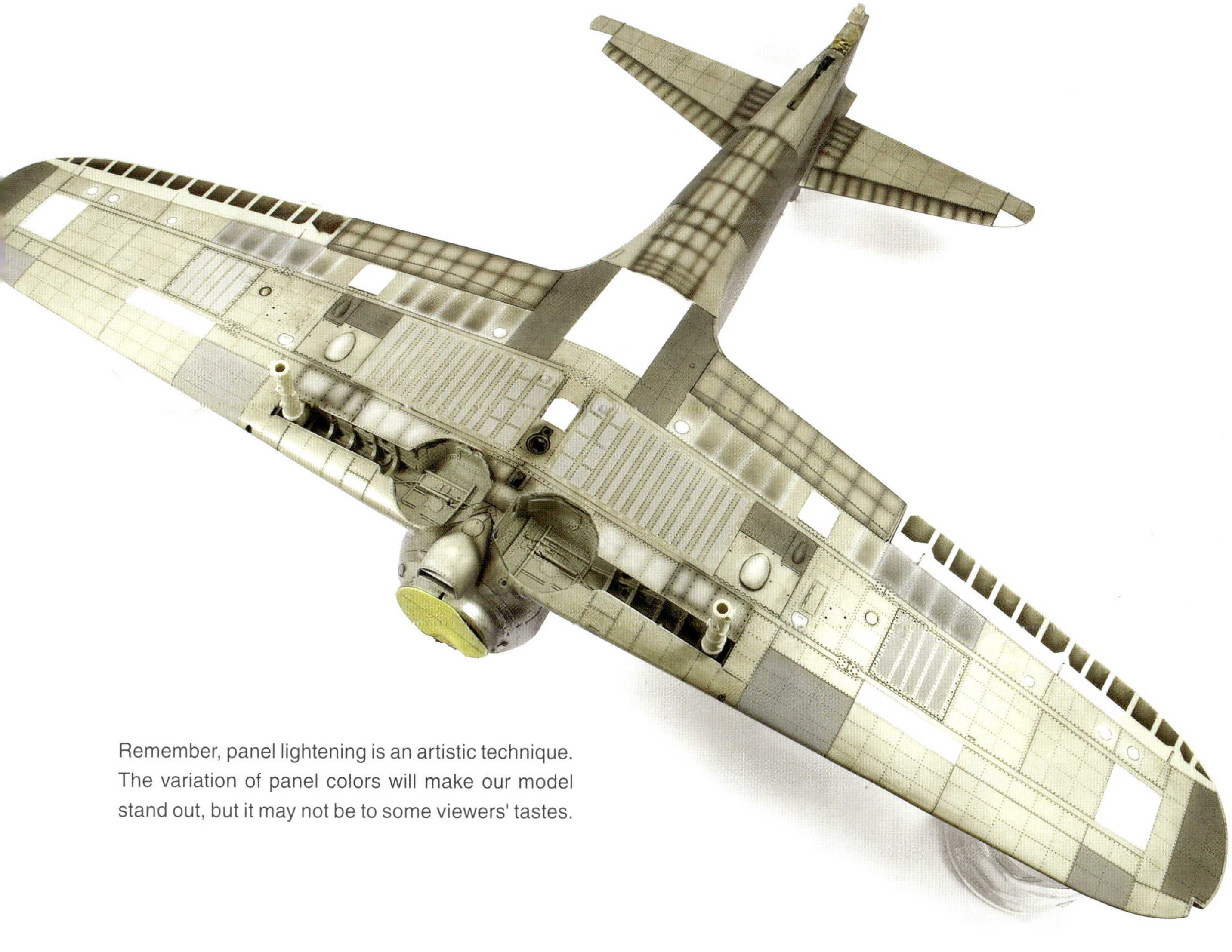

Remember, panel lightening is an artistic technique. The variation of panel colors will make our model stand out, but it may not be to some viewers' tastes.

Pre-shading

As you've seen in previous sections concerning painting and finishing techniques, you're trying to give your model dimension. Pre-shading can help do that. By pre-shading, you outline the shadow areas before applying color modulation. The best choice for this work is black or, for some variation, dark blue or dark green.

Depending on your model's scale and type, this technique can be anything from simple to demanding, requiring dexterity and precision. Easy or difficult as it may be, pre-shading offers effective results. Although pre-shading is mostly associated with aircraft modeling, as with all other finishing techniques, it can be applied to all kinds of models.

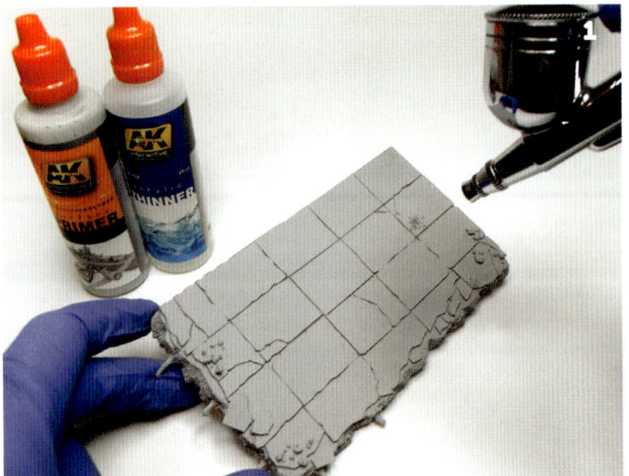

1. Some modelers begin pre-shading after priming. Others use different primer colors for pre-shading. Your experience will dictate what works best for you, because what's important is the final result.
To pre-shade, preferably start with a thin coat of white or gray primer. You can work with different color primers, so long as it isn't black. However, this will effect the contrast when you apply the base coat.

2. Next, airbrush the pre-shading. In this example, the scoring and cracks in the concrete get a treatment of black paint. This operation is delicate and demands precision—you want your lines as thin and straight as possible, but they don't have to be perfect.

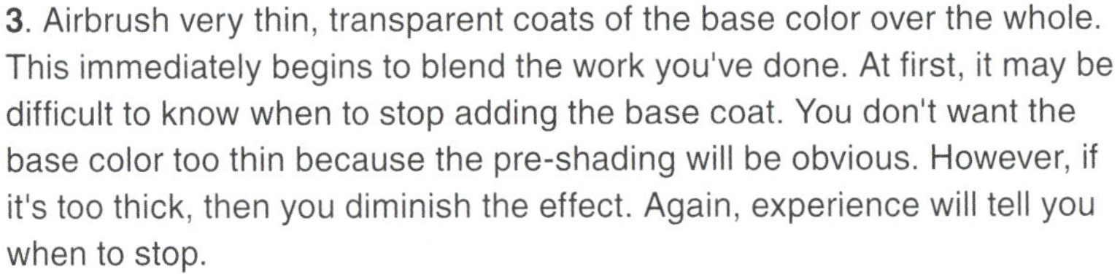

3. Airbrush very thin, transparent coats of the base color over the whole. This immediately begins to blend the work you've done. At first, it may be difficult to know when to stop adding the base coat. You don't want the base color too thin because the pre-shading will be obvious. However, if it's too thick, then you diminish the effect. Again, experience will tell you when to stop.

As you can see, the pre-shading adds dimension to the diorama base.

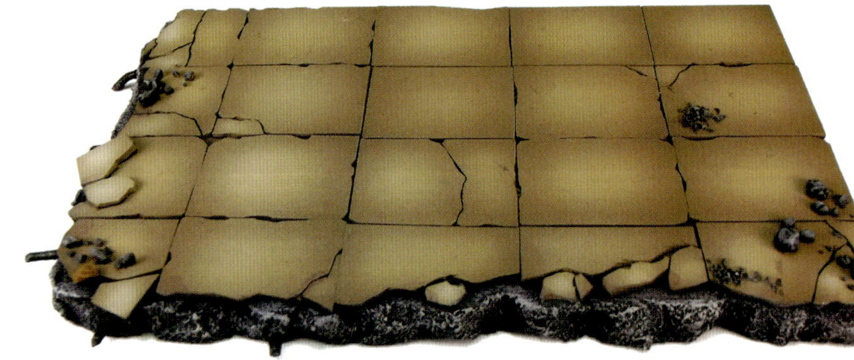

The same procedure is widely used in aviation modeling to highlight the panels and rivets. It is meticulous and precise work, especially in small scales.

Some modelers don't start their work with pre-shading, but instead apply it as an intermediate stage. In this case, after the surface has been painted with the base color, airbrush the pre-shading with either black or a dark shade of the base color.

Then airbrush extremely thin layers of the base color over the whole until the shadows soften to the desired look.
Pre-shading alone, in most cases, does not offer enough color modulation. Other lighting effects must follow before you're finished painting.

Post-shading

Post-shading, or simply shading, is about airbrushing the shadows to create more contrast directly on the painted surface of the model. Ideally it should be done with a darker version of the base color or with colors related to black (blue-black, brown-black, etc.)

Use an airbrush with the pressure set low and a 9:1 thinner-to-paint solution for shading. As with all light techniques, use thin coats and let each dry before adding another. Be subtle and build up the effect by overlapping layers. Use a critical eye on your work until the shading it just right.

Shading can be done freehand, or using masks. Tamiya, Gunze, and AK Interactive acrylic colors are great for shading because they behave predictably when highly diluted. To get thinner lines, remove the nozzle cap from the airbrush and reduce the pressure. To be successful painting like this, you should be comfortable with your airbrush, thinning paint, and adjusting air pressure. If you use a mask, aim the paint spray exactly at the edge of the mask. When you remove it, you will be left with a fine shadow.

Paint bursts must be short and rapid when painting this close to the model. Long bursts run the risk of overspray, spurting, and other mistakes.

Post-shading can be done with lighter colors too. You can enhance the effect by painting a light line next to the shadow, increasing the contrast and sense of dimension. Yet, in this expanded form, the resulting effect is extreme and must be applied sparingly. Normally, shadows must be darker than the base color, hence the name of the technique.

In the following example you see how to use heavily diluted paint to simulate the ribbing on a biplane wing. The wing ribs must be carefully masked and successive coats of shading are applied.

Dry-brushing

Dry-brushing highlights small edges and details on your model that would be almost impossible to paint with a brush. Historically, it's one of the two oldest finishing techniques—washes being the other. Dry-brushing can be used on any model.

Although it doesn't seem so, dry-brushing is a complicated technique that needs practice and experience to be effective. If rushed or overused, dry-brushing can make your model look frosted and unrealistic.

Dry-brushing requires patience and control. You'll need a flat brush, the paint you want to dry-brush (enamels, acrylics, or oils can be used), and a piece of paper towel or a spare cloth.

We dip the brush in the paint and immediately rub the bristles on the paper towel. Unload the paint until almost no color comes off the brush. Then brush the edges and raised details on the model with quick, soft strokes.

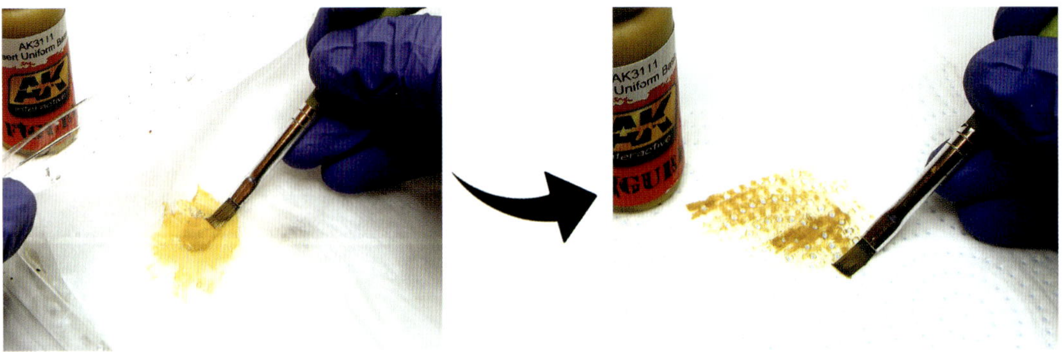

Below is a small scene with sand and rocks. First, prime with gray. Then, brush on the highlights and shadows on both the ground and the rocks.

Concentrate on the main elements without going into details. Now, you'll dry-brush.

1. Begin with the sand. With a color lighter that the base but related to it, load the brush as described earlier, wipe off the majority of the paint, and softly and quickly brush the sand. The result in this first step is hardly noticed.

2. Using the next lighter color, repeat the process, concentrating on the more salient points and the edges. The effect it is now becoming more evident.

3. To finish the work on the ground, add white to the color from the previous step and repeat the process again. By now, it may seem that the result is overdone, but successive weathering layers will soften the effect.

Use the process on the other features on the base. Everything has more depth and richness of color that it previously lacked.

Combine dry-brushing with other techniques that unify all the elements of your model to achieve greater realism.

TIPS

- If the area to be treated is highlighted after the first three brush strokes, then your brush is not dry enough. Always use a flat brush for dry-brushing.

- Wet patches or smears means the brush is not dry enough.

- When you think you've unloaded the brush enough, think again! Make a couple more swipes.

Masking

Masking is the process with which you prevent paint from spreading to areas you've already painted. A mask can be made with many different materials, such as paper, masking tape, a Post-it note, or even a piece of metal foil. In any case the masking procedure is the same.

To protect large areas, you can carefully border the area with masking tape and cover the rest with paper.

When airbrushing over a mask, direct the airbrush band perpendicular to the mask, so no paint is accumulated on the edge of the mask or seeps beneath it.

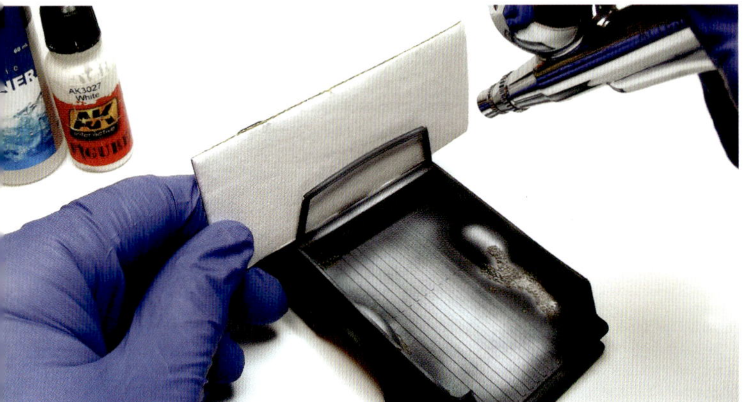

1. TEMPORARY MASKS

These masks are used to border a different color area or surfaces that must not receive paint. They're called temporary masks because you can put them on and remove them on the spot. For this type of masking, you can use any type of paper or cardboard that is rigid enough so you can hold the mask without distortion while you airbrush. Cut these masks to the desired shape and hold in position with one hand. Try to direct the paint from the masked area toward the target area.

2. LIQUID MASKS

Purpose-made masking liquids can be brushed directly onto areas you want to mask. When cured, they turn rubbery and can be easily removed after you've finished painting. Liquid masks are best used to cover small areas, details, clear parts, or to make chipping effects. Remove the mask with your fingers, a toothpick, or a pencil eraser. Be gentle and deliberate so you don't damage the paint.

3. MASKING PUTTY

Masking putty temporarily adheres to the surface of your model and can be removed without leaving any residue. Being elastic, it has the ability to adapt to any shape or surface. It also has the advantage of being reusable. Masking putty has a multitude of uses, but one of the most common is for airbrushing camouflage patterns.

4. MASKING TAPE

Masking tape is the most common form of mask and widely used for precision jobs. You can find it in a range of widths and thicknesses, and even in sheets that can be cut to custom shapes.

Remember, it's always better to work with the right tools than to improvise. So, draw the outline of a mask with the help of a ruler and a pencil. Next, cut these parts with a sharp No. 11 blade. Finally, place the mask on your model, taking care to cover only the areas you want to protect. Check that the masks are secure without any gaps or bubbles along the edges. After the mask is correctly positioned, you can apply the paint and desired effects without the risk of spoiling previous work. In general, masking tape has the correct amount of adhesive to stick without damaging the paint beneath it when removed. Still, the less time the mask is left in place, the better.

Remove the mask when you're finished painting.

This type of mask is ideal for straight lines.

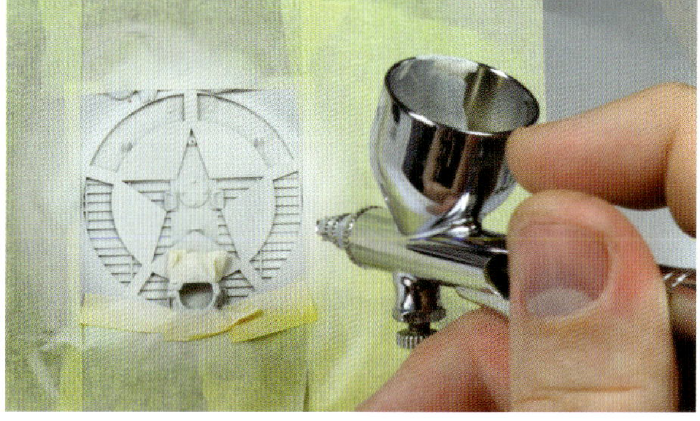

5. MASKING TEMPLATES OR STENCILS

Masking templates and stencils require some experience. They are mostly used for patterns and shapes with a soft edge finish or without much detail. Rigid materials can be used for masking templates, such as cardboard or photo-etched metal frets.

Special templates with different shapes, emblems, or writing can be found in art and hobby supply stores for this purpose.

TIPS

- Although athletes' tape, plastic wrap, plasticine, and Blu-Tack can be used for masking, it's better to use materials designed for the purpose.

- Cover as much of the masked area as possible so paint doesn't reach even the most remote parts.

Filters

A filter is an ultra-thin layer of paint or other finishing medium that, when applied, changes the shade of the underlying color. You can achieve a more realistic finish, adding depth and richness to your work. Imagine that you lay colored transparencies over a photograph. Filters offer exactly the same visual effect. You can't get this effect by simply mixing colors.

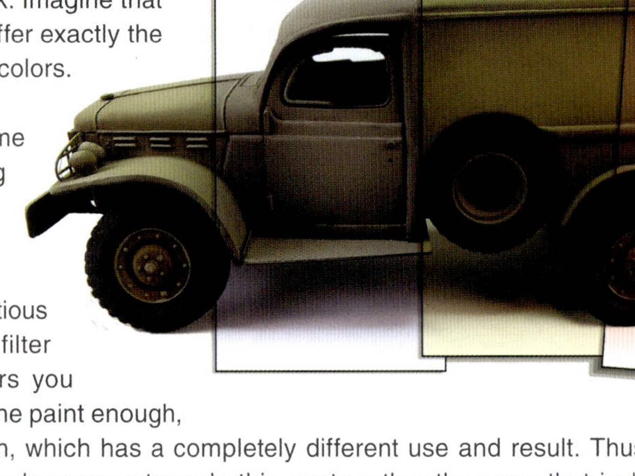

You can use filters on any type of model. Some filters come ready to use, like paint. You can also make your own, mixing 10% enamel or artist oil with 90% solvent.

Note that you must be cautious with thinning a custom filter and the number of layers you brush on. If you don't thin the paint enough, you'll end up with a wash, which has a completely different use and result. Thus, it's always preferable to apply more extremely thin coats rather than one that isn't thinned enough to properly do the job.

Filters are the next step after color modulation. First, airbrush a couple of coats of satin or gloss varnish to protect your previous work.

Why not clear flat varnish? Colors with a flat finish tend to be rough and make the filter appear more like a stain than a separate layer.

On gloss surfaces, a filter will adhere properly, creating the same effect of a filter placed over a camera lens.

Now, let's apply two filters: one light green, and another darker with a brown tone.

As mentioned above, filters can be used on any type of model. They are easy to apply and provide interesting results.

Successful filter application means subtly dying the underlying paint colors. You can thin a filter with solvent to subdue its effect if necessary, especially when working with light colors. Apply filters using a soft, flat paintbrush, the thicker the better.

Wet the brush, removing the excess on a piece of paper, then apply onto the surface. In this case, the whole vehicle will be covered. Wait two hours before brushing on another coat, even if it's of the same filter.

The filter effect will not be visible until it has thoroughly dried. Apart from the color variation it adds to your model, filters will help blend gradients, brush strokes, and minor discolorations.

Filters are especially beneficial when used over camouflage schemes, either on uniforms or vehicles. Modelers of any skill level can easily and quickly apply filters.

These three areas on the fuselage of this plane are bordered simply using a filter. Filters can be used to distinguish areas of different materials, or surfaces where paints weather differently.

TIPS

- Keep a paintbrush reserved only for filters.
- Filters shouldn't be overdone.
- Use filters to change monotonous and uniform surfaces.

Washes

Use washes to pick out details that might otherwise go unnoticed. The basic idea is to place a dark color into the scribed detail of a model in order to outline those details.

While you can buy washes, they're easy enough to prepare on your own with enamel or artist oils. When making or selecting a wash, the most important factor you need to consider is the model's base color. You want to use a proper color that will not be too stark.

When painting figures, washes can help shade difficult-to-reach spots. Combined with the dry-brushing, you get more contrast and a dimensional effect for ground, hair, fur, feathers, and the like. Nevertheless, a wash isn't a technique largely effective with figures, so it must be applied selectively.

Washes can replicate accumulated dirt, grease, and light dust in recesses and lines of any vehicle. You can apply a general wash, allowing the color to freely flow and accumulate where it may, or use a pinwash, which is a more targeted and controlled approach.

In this sequence and on the following page, you can see the notable difference between a model without a wash and the depth and detail a wash reveals.

After washes are applied, we can continue with any other finish effect.

METHOD

As in the case of filters, work on a satin or gloss finish. You can airbrush a flat coat after the wash is down.

• With a brush, apply the wash in all recesses and lines, allowing the wash to flow and accumulate into them. It isn't necessary to be very precise, because a wash can easily be wiped away. Work in small areas to minimize fatigue and to give the wash a chance to dry.

• With the help of the brush and an odorless solvent for enamels, you'll blend the effect so that it merges with the paintwork. Remember to direct the wash toward the recesses you wish to highlight.

• Moisten a brush with mineral spirits and remove any excess wash, cleaning areas that have been stained. If your washes are acrylic, soak the brush in water. You have to work fast and smart with acrylic washes because they dry quickly. Enamel or oil paint washes are more forgiving.

• When finished, leave the washes to dry for two hours. If you want, you can then airbrush on a flat varnish.

TIP

- Makeup brushes, cotton pads, and cotton swabs make perfect disposable tools for cleaning up excess washes.

Panel lining

Scale models of airplanes, cars, tanks, trains, and even concrete slabs and doors all have a common characteristic: They consist of separate parts usually bordered by fine or larger lines or scoring. Generally termed panels, the bordering lines between parts, no matter how well finished and flush, accumulate dirt due to time, operation, and wear.

If we paint our models ignoring the panel lines, the result may look artificial and flat. **Panel lining** is the technique used to treat these lines and highlight them.

After panel lining, your work will gain depth and volume, thus enhancing realism. Panel lining is always done on a model that has already been painted, with decals and other markings already placed. A coat of gloss or satin varnish is a must to allow the paint to flow well. There are several ways to create the effect, the most common are explained below.

USING PANELINERS

Thinned enamel paint, Paneliners come in a wide range of colors. Shake well before using. With a fine-tipped brush, flow the Paneliner to the panel line.

Wait 40 minutes, then remove any excess Paneliner with a flat paintbrush moistened with enamel thinner.

Always pull the brush in the direction dirt or grime would normally streak on the model. For instance, from nose to tail on an airplane.

We can repeat the process as many times as necessary until the desired effect is achieved. Some Paneliner colors work better than others depending upon the color of the model. In general, it's good to have at least one tone for sandy colors, and another for greens and blues.

PANEL LINING WITH ARTIST OILS

Pick a color similar to the base color. Darken with 50% black. Dilute the mix with turpentine until you get a consistency similar to milk—thinner if you prefer. In any case, our mixture must be liquid enough to flow into the recesses, yet retain enough pigment to darken them.

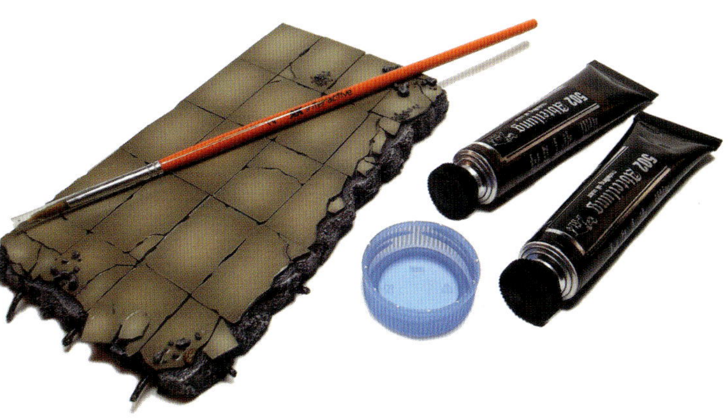

Lay down a coat of gloss varnish. After the varnish is dry, brush the artist oils into all joints between panels, the cracks in the concrete, and on the accumulated debris. Work carefully to avoid staining the surface.

Let the artist oils dry for 30 minutes.

When the oil paint is dry, remove the excess liner using turpentine or mineral spirits. You can clean large areas with paper towel. Use a makeup sponge or cotton swap for smaller, hard-to-reach places.

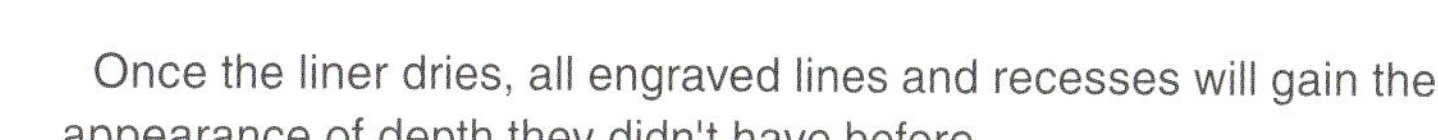

Once the liner dries, all engraved lines and recesses will gain the appearance of depth they didn't have before.

If a line or panel is not covered well, repeat the process. Sometimes it's a good idea reapply the panel lining a second time.

TIPS

- Wear nitrile gloves (some people are allergic to latex) during the panel lining process to safely handle the model.
- When working with artist oils, stir the mixture often because the pigment tends to deposit in the bottom of the container.
- If you use too much solvent during the cleanup stage, it can minimize or remove the effect.

Mapping and dot filters

When painting any type of scale model our base colors are usually solid and flat. The trouble is that in reality, colors are seldom flat and uniform. Whether we realize it or not, colors usually degrade into numerous tones and shades. The reason for this are many—bad paint, the vagaries of use, harsh weather. Whatever the reasons, colors eventually acquire a multitude of wear effects that affect the overall tone.

Two of the most effective techniques to replicate this degradation in paint color are **mapping** and **dot filters**. The objective is to create a semitransparent layer of different shades and tones over the underlying colors. Mapping and dot filters are usually applied near the end of a build, after decals have been placed.

Mapping is named after the distinctive shape of the spots you create, which, from a distance, can look like continents on a map. In this example, you will see how to apply mapping with an airbrush. You can map with a brush too.

Although our vehicle has already undergone weathering, it still looks flat. Your goal is to try make the paint look old and dirty, ready for further weathering effects.

When it comes to dot filters, understand that you will work with colors applied to vertical surfaces differently than you will with colors applied to horizontal surfaces. Work in small areas, finish the filter, then move on to the next area. This will create an uneven filter that will appear more realistic.

VERTICAL SURFACES

1. Apply dots of artist oils with a brush on the working area. Keeping in mind that the light colors must be up high and the darker colors toward the bottom.

2. Wet a second brush with solvent and remove the excess on a paper. Brush the paint from top to bottom—only in that direction!—blending the colors together.

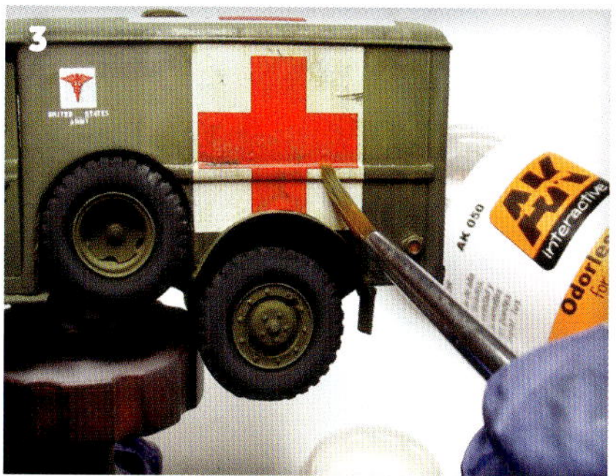

3. Repeat the previous step as many times as necessary to get the desired effect. The more you work with the solvent, the less noticeable the effect will be.

HORIZONTAL SURFACES

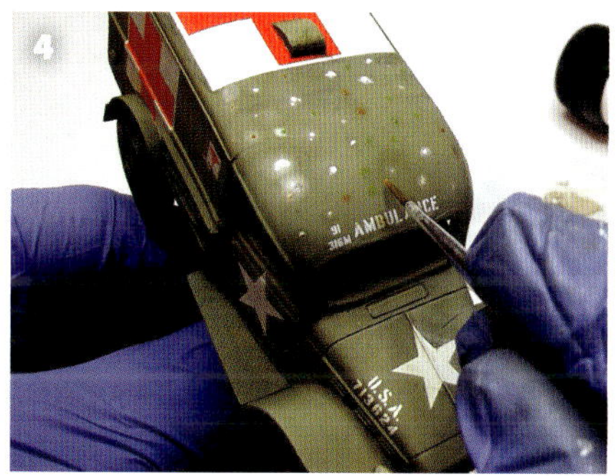

4. Apply dots of artist oils with a brush on the working area. Use only light colors because it's an area that receives a lot of light.

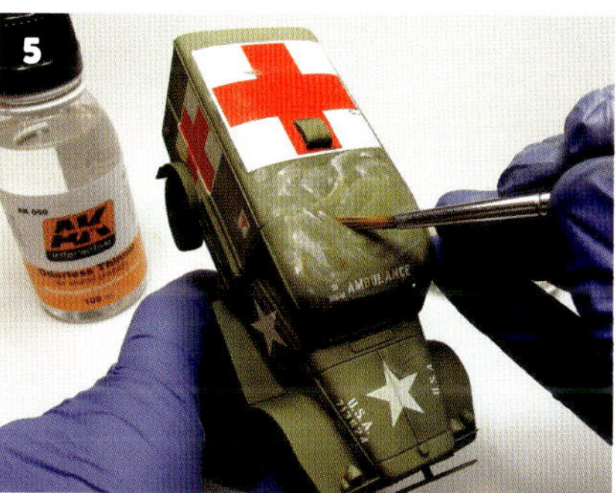

5. With a second brush moistened in solvent, apply soft, circular brush strokes, almost as if you were drawing the shape of a cloud.

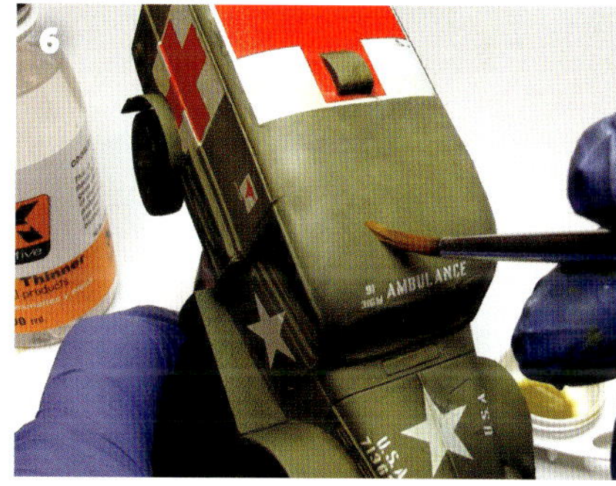

6. As with the vertical surfaces, repeat as many times as necessary to get the desired effect without completely removing the layer of artist oils.

Let the model dry overnight, and then airbrush a coat of varnish to protect your work before you continue weathering.

TIP

- Artist oils take longer to dry than other paints. Make sure to wait at least 24 hours after using them before proceeding.

Applying textures with a brush

There are times, particularly on large-scale models, that something seems to be missing. It all looks wonderful, but there's just something you can't quite put your finger on. And then it occurs to you that an item of clothing or a decoration really should have some sort of texture.

As you'll see in an upcoming section, you can achieve respectable texture results using a sponge. However, there are some models that won't work on because they require greater control over the outcome. A paintbrush allows you just this sort of control, creating **line** and **stipple** textures. Let's take a look at two leather surfaces and how you might utilize these techniques to create different textures on similar surfaces.

LINE

Old leather acquires cracks and splits and worn spots. These mostly appear near the areas of friction and folds. Below, you can see how to create these textures using a lined pattern. You can also adopt this technique to create textures for coarsely woven cloth and corduroy, as well as many other materials that have seen a lot of use.

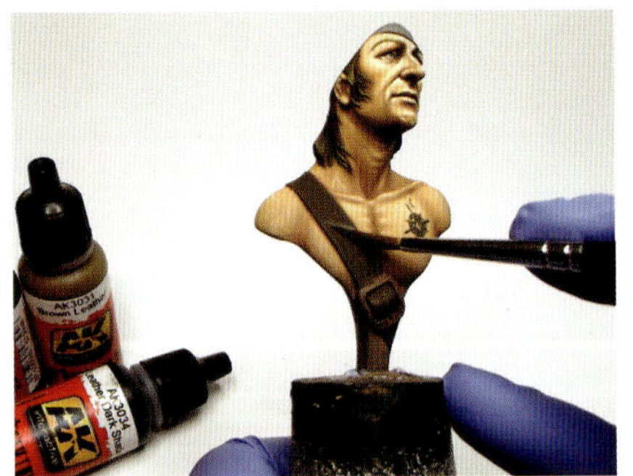

Intending to add texture to the belt over the pirate's shoulder, start like you normally would by brushing on the base color, highlights and shadows, and color modulation.

To get the effect of old, broken leather, take your highlight color and paint small lines along the edges. Also, paint scratches at odd angles on the face of the strap.

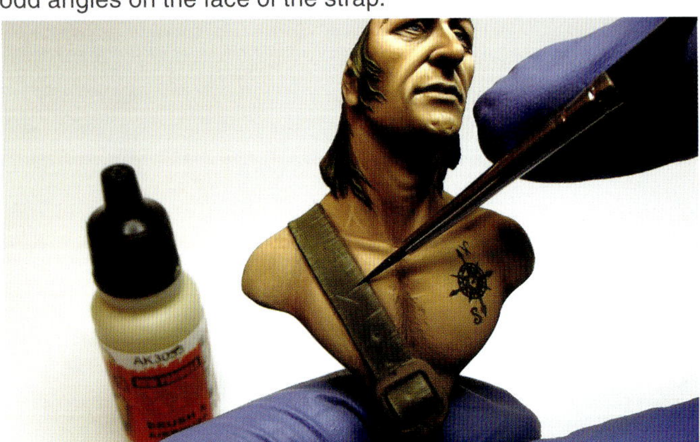

With a very dark shade of the base color (or black), make some of the lines along the edge deeper and add a couple small, random stripes, over or above the lighter lines.

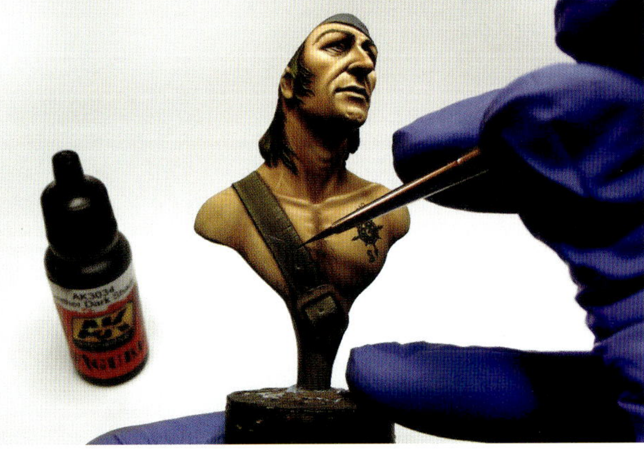

STIPPLE

Another common pattern found in many textured objects is that of dense uniform dots called a stipple effect or stippling. The technique creates a porous or aged look. You make small dots with the tip of a brush, and include different shades of paint for variation. This way, you can replicate many different materials like stone, old fabrics, and leather. As an example, let's simulate the aged and worn leather of the pirate's hat.

First, as with the strap, paint the base coat and do any color modulation you have in mind. In this case, using an airbrush, the hat is pre-shaded before spraying on the base coat.

Then, making tiny dots with a fine-tipped brush, add the extreme highlights on the areas where light is most likely to hit. Gradients lead into the narrow areas where there's shadow near the highlights.

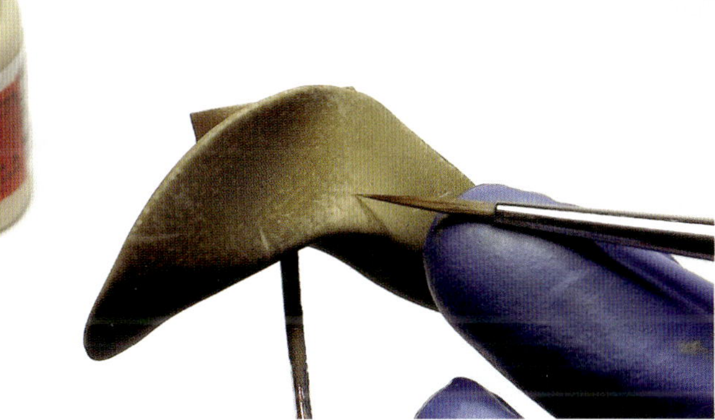

Use the same process with the shadows. On some areas add some of complementary color (remember the color wheel!) to the base color to add even more depth and variety.

At this point, you can consider painting finished, but you can also add a glaze or a filter. This helps blend and unify both the line and stipple effects, making the result even more natural.

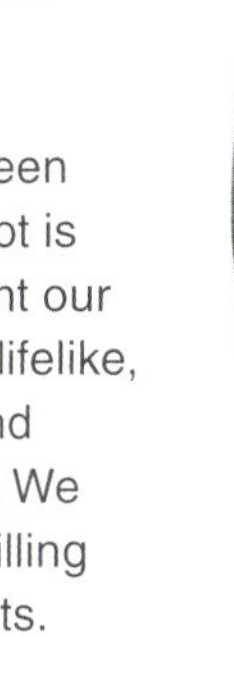

The difference between adding textures or not is remarkable. If we want our creations to look more lifelike, we have to stretch and employ such techniques. We must be bold and always willing to apply new skills and effects.

TIP

- When we apply stippling, use an old brush and randomly tap the surface. On areas of highlights, use a brush with very fine tip for better control.

What does weathering mean?

By now, you've selected and gathered all our materials on the workbench; you've researched, prepared and assembled our project and applied the base color. You added the necessary light and color modulation, and you're ready to take the next step toward an eye-catching, life-like model. Still, you look at your work and feel something's missing. So, what's wrong? If you've followed all the steps so far, why doesn't the final result, satisfy you? Well, it's because you still haven't added the weather-, time-, use-, or battle-effects that will make your model unique. In most cases, you don't want your projects to look as if they have just left the factory. Nor to look like toys. The way to resolve these issues is to apply a series of techniques that allow you to replicate the patina, exposure to nature, traces of use and battle scars on your models. These techniques, as a process, are called "weathering," in scale modeling.

Each of the weathering techniques can be applied individually, in combination with, or complementary to others, depending on the result you want to achieve. For weathering, you can use any type of material and tool, as long as you know what to do with it and how to do it. All types of colors, from acrylics to oils, knives and sandpaper, pigments and pastels can be used to make your model look old and damaged. There is a dedicated range of products available created exclusively for weathering, so one can buy a product for a specific color, or weathering effect. These products will greatly improve your work. Some caution is required when using them, since they're quite "aggressive" materials (enamels, lacquers etc.), so it's highly recommended that your paintwork be protected with several layers of varnish before any weathering is applied.

After a certain period of time, you'll notice that an object left outdoors to endure the elements will begin to suffer from a number of different weathering effects. By looking closely, you can see the kinds of effects and extent of their ravages you want to model. Rather than follow a weathering script, imagine what you want to see; what you want to do. One of the first things you'll notice is how the paint cracks and chips, gradually revealing the material beneath. Chipping is very common among military vehicles, heavy machinery, and metallic surfaces in general. Whenever you can, get out and see how materials age in nature, or find photos to help you simulate realistic weathering.

When an object is left outdoors, dirt inevitably accumulates on it. When it rains, dirt mixes with water and flows down, creating streaks. It's very common on any object that's ever been rained on.

Rust is a thin layer of corrosion created on steel and iron surfaces when in direct contact with water or exposed to moisture. Rust can also affect and stain other materials that come into contact with the metal. It's very common on vehicles.

Dirt tends to accumulate in recesses and on horizontal surfaces. This is called a dirt deposit. Water creates mud, and when vehicles move through it, they get splashed. This is yet another common weathering affect.

Dust is another element that will get all over and in vehicles, or people for that matter. In scale modeling, dusting is a basic weathering effect that can be overdone. Always use logic and check your references before applying dust to your model. A ruined clock tower in a city will not have the same amount of dust as a tank crossing the desert.

All these effects and many more will help you make your model look realistic. The way and the order of applying and combining these effects depends on logic and your vision for the final result. Only through experience can you know what works best for you.

TIPS

- Weathering must be coherent and logical. A piece of metal does not age the same as a piece of wood, nor does it react the same when in contact with water. Always study references before modeling weathering in a particular environment.

- When it comes to weathering, remember, in many cases less is more. Show restraint.

Chipping

Chipping exposes parts of the primer or the material under the paint when the latter deteriorates due to wear. When possible, use reference photos to help you keep your chipping under control, because it can quickly get out of hand. Chip patterns may look random, but they're anything but. Chipping replicates the loss of paint from one or all subsequent layers of a vehicle or any paint-covered object. Depending on the underlying layer, chips might appear either darker or lighter than the uppermost layer. For chips representing bare metal, a chocolate brown color is recommended.

Chipping usually appears in areas prone to use and damage, such as fenders and around hatches. If the chipping is deep and has reached the metal, it's possible that it has rusted. Perhaps rain has streaked the rust.

On larger scales or in very large scratches, you can outline the bottom or top of the chipped area to highlight it. This representation of light will make the effect richer, more realistic, and more interesting.

Always keep in mind that there are certain metals, such as aluminum, that do not rust.

With such metals, we avoid the usual rusty tones, since the chipping under the paint will only expose the bare metal.

When chipping, make sure to always create asymmetric edges (no rounded edges) and irregular patterns. Doing so will make the effect look natural. Also, make differently sized chips with scratches of irregular shape. Additionally, always keep in mind the underlying layers of color and the kind of metal found beneath. This way you can effectively select the colors you'll use for each successive coat.

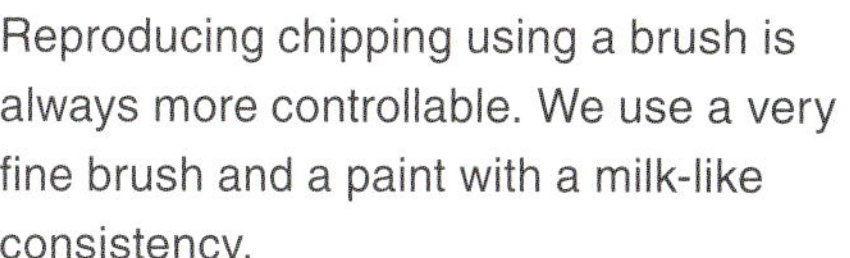

Reproducing chipping using a brush is always more controllable. We use a very fine brush and a paint with a milk-like consistency.

Not all parts of a vehicle or an object are chipped the same way, because raw materials react differently to environmental conditions, altering the type and look of the chipping. Paint chips from wood differently than from metal because of its grain and porous structure. Sometimes, the chipping doesn't reach the material surface, exposing an underlying layer of paint or primer.

Chipping over real wood.

Chipping over simulated wood.

Chipping and superficial scratches.

Watercolor and graphite pencils

Watercolor and graphite pencils are another example of materials adapted to modeling from fine arts. They're both manufactured in a pencil form, so they offer the advantage of being easy to use, while providing a very fine tip to work with, requiring only a regular pencil-sharpener to maintain the point.

Watercolor pencils are colored pencils that are water soluble and can create watercolor effects with a brush and water. They can be used both in dry and wet form. The major disadvantage is that they're not cheap, and it's difficult to find specific colors individually. You usually have to buy a set that will certainly contain colors you may never use.

The only limit to how you might use these pencils is your imagination. They can be used to outline, highlight tiny details, replicate chipping and scratches, and myriad other effects.

Besides being a versatile product, they're easy to apply and control. It's also easy to correct any mistake: Just remove the error with a cotton swab moistened with water.

Watercolor pencils don't require a varnished surface to be applied. They are mild and do not react with any of the finishing products normally used for scale modeling.

Graphite is the core material of any normal pencil. Pencils are common and inexpensive. Although regular pencils can be sharpened to an almost deadly point with a regular pencil sharpener, a mechanical pencil is a better choice. Their fine tips offer more precision, and you can get a variety of graphite hardnesses to suit your needs.

With the sides of a mechanical pencil's graphite, you can highlight edges to make them look like polished metal. A 2B or 3B graphite works best.

All you need is a clean white eraser to correct any mistakes. Be careful how much pressure you apply, because extensive rubbing can damage your paint job.

Graphite artist pencils are more limited than the watercolor pencils due to their metallic effect. Yet, they can be used for a multitude of effects, such as chipping, scratching, and so on. These pencils can only be used dry. Some modelers also use them to color panel lines.

There are a variety of conventional graphite pencils made specifically for modeling use, with colors like chalk white and brown—good for creating chalk marks on vehicles and chipping, respectively.

TIP

- Apply pencils after a coat of varnish.

Hairspray technique

The hairspray technique appeared long before adequate purpose-made weathering products appeared on the market. This technique allows us to replicate a poor quality finish on a vehicle or other painted object. It easily re-creates missing paint chunks, revealing the color or material lying beneath. Hairspray has some drawbacks in comparison to available modeling mediums. It's sticky and consistency varies depending on brand, making it difficult to standardize application. Often it isn't easy to work with in a controlled manner and the result may not always be what you desired. In general, the intensity and shape of the chips depends on the amount of hairspray applied and the drying time you allowed for the last coat of paint.

The hairspray technique is good for faded, chipped, and damaged paintwork typical of vehicles and objects subjected to fatigue, heavy use, nature, and battle.

It can also be used to simulate faded and worn winter camouflage applied to tanks during World War II. Such coats were generally of very poor quality, mostly based on water or fuel, and prone to fade and wear.

PAINT OF BAD QUALITY

To simulate temporary winter camouflage, apply this technique to any vehicle that has been over-painted white with bad paint.

• First, paint your model with its final finish as if it was not to be covered in white camouflage. Apply all color and light modulation, and even the decals and other weathering, to model wear before the white camouflage went on.

• Apply hairspray to the model, up to three times with a spacing of 5 minutes between each coat. This may vary, depending on the hairspray you use and the effect you want to achieve. The more coats of hairspray, the more chipped the result.

• 15 minutes after the last coat of hairspray, airbrush on the white paint in a thin layer. Coverage should be uneven and spotty.

• Lastly, remove part of the white paint with an old, stiff-bristled flat brush or piece of sponge and warm water. Wet the target areas with another brush and then gently rub with the flat brush or sponge. The white paint should come off easily. You can also simulate scratches with the tip of a moistened toothpick.

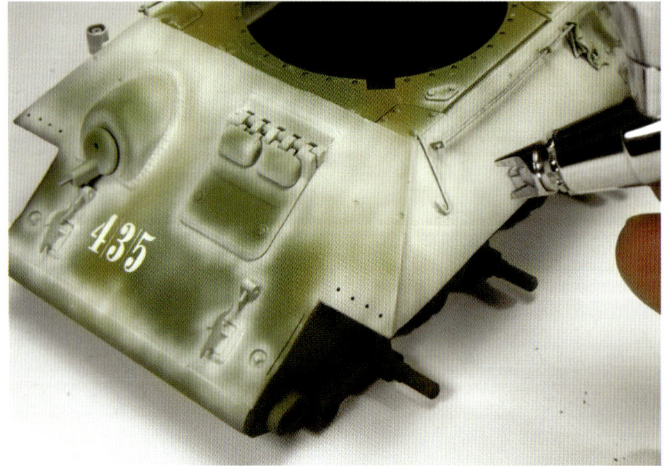

CHIPPING EFFECT

Next, using the hairspray technique, you'll learn how to simulate areas that have received damage and the paint has chipped, exposing either the primer or the bare metal beneath. To make it more eye-catching, let's make a rusty inner layer so it will look like the chipped part has been exposed to the elements for a while. As usual, first lay down base coat of red-brown for an oxidized look. This layer will be the one visible when the effect is applied.

- Spray a generous coat of hairspray and let it dry for 15 minutes. You can apply it directly from the spray can or transfer it to your airbrush.

- Next, airbrush a coat of paint over the hairspray. For this fuel drum, a sandy yellow color will do nicely.

- Once the yellow has dried, moisten the fuel drum with water. Depending on the effect you want to achieve, you can wet either the whole surface or different areas of it using a flat brush.

- Feel free to use different tools to chip the yellow paint: a cotton swab, file, knife point, toothbrush, sandpaper, or toothpick. Each will create a different result; use more than one.

- After the technique is applied, allow the fuel drum to dry thoroughly. Lastly, spray a layer of varnish to secure and protect the effect.

TIP

- Sometimes the quality of the hairspray works opposite in scale modelling. The lower the quality, the better it works on our model. The fixing properties are what make the difference.

Chipping with acrylic medium

One of the fulfilling challenges in scale modelling is to represent the effects of time, use, and nature on a model. There are many ways to do so on our work. Chipping ranks as one of the easiest and best ways to accomplish this. In this section you'll see possibly the most popular chipping technique and learn how to do it: acrylic chipping mediums.

There are two types of chipping mediums available: one that will give us large chips and a second that provides smaller chips and scratches. Bear in mind that these liquids are acrylic and activated by humidity, so the coat of paint that goes over them has to be acrylic. All previous coats can be of any other type of paint. They don't matter. And in any case, they won't be affected by the medium's agents.

The method for using acrylic chipping fluid is quite simple. They are made to create an intermediate layer between the surface color (the visible color) and the colors that lie beneath. Decide how aggressive you want the chipping to be before applying the chipping medium.

1. First, apply the base color. This is, the color that will be revealed through the chipping. This model has been modulated with several rust colors.

2. After the base coat has thoroughly dried, paint on the chipping medium with a brush. When it's dry, we airbrush the surface color—the one to be chipped. Remember this coat must be acrylic paint.

3. Let the surface coat dry completely. This will later affect the size of the chips.

4. With all the layers dry, moisten the surface with water, activating the chipping medium. Gently rub the surface with a stiff-bristled brush.

5. Use different tools to make the chips and scratches. A toothpick, file, or tip of a knife blade all make excellent and unusual marks.

6. The process can be repeated with a new layer of paint, simulating a repainted surface. The resulting chipping, in two colors, will look even more interesting.

7. The last step is to make the chipping appear beneath the new layer. Again, moisten the surface with water and rub it with a flat brush. Now the chips reveal randomly both the underlying color and the rust layer.

TIPS

- Study how different materials age before chipping. Use logic: A wooden object will not be rusty.

- The chipping effect increases the more chipping medium you apply—more medium, more chipping.

Sponge technique

When replicating worn paint schemes and textures, we sometimes need to simulate patterns consisting of tiny spots with random shapes and sizes.

Such a time-consuming and arduous task would be practically impossible with a paintbrush. The sponge technique is an ideal option. Using sponges or foam of different porosity, you can achieve several patterns of different density and size.

Packing and protective foam found in shipping boxes are wonderful for this technique, but any kind of sponge or foam can be used. There are even sponges in the form of brushes available, their tips made specifically for this technique. In any case, to get an uneven and random result, slightly deform the sponge before using it to avoid straight edges.

Overall, using sponges is very simple. Pick your colors and pour a generous quantity of paint on a palette or piece of cardboard. Cut a piece of sponge to an appropriate size for the object you're working on, and slightly deform its surface to get a more uneven pattern. Dip the sponge in the paint and unload most of it on a paper towel, the same way you would with dry-brushing. When there's almost no color left on the sponge, gently tap it on the areas you want to treat, taking into account the density of the texture you want to simulate, and of course, the scale of your model. Repeat the process as many times as needed to get the desired effect.

Four different brownish tones have been applied to these wheels, getting a very realistic rust effect.

In this example, you can see how irregular and random the paint looks on the surface. The patterns were made quickly and easily. It would be almost impossible to get such a result with a brush.

Multiple sponges with different densities can be combined. The denser sponge should be used on smaller areas, and the looser one for larger surfaces.

TIPS

- Don't thin your paint when using the sponge technique.
- As with dry-brushing, it's always better to apply too little than too much. Repeat the process more than once to get the effect just right.

Salt technique

As its name indicates, the salt technique uses salt to mask a layer of paint to enable us apply another on top. When the salt is removed, the underlying layer is exposed. With this technique you can create chips, rust, minor wear, or gravel and ammunition impacts.

You use a material easily found in any house: salt! Its grain will determine the size of the resultant chipping. You'll also need a cup of water, a paintbrush with stiff bristles, and a toothbrush. To begin, apply a uniform layer of primer to your model. Although the example is gray, a rust-colored primer would greatly accelerate the process. If there are areas of your model on which you do not want the effect to be applied, you must effectively cover it before starting.

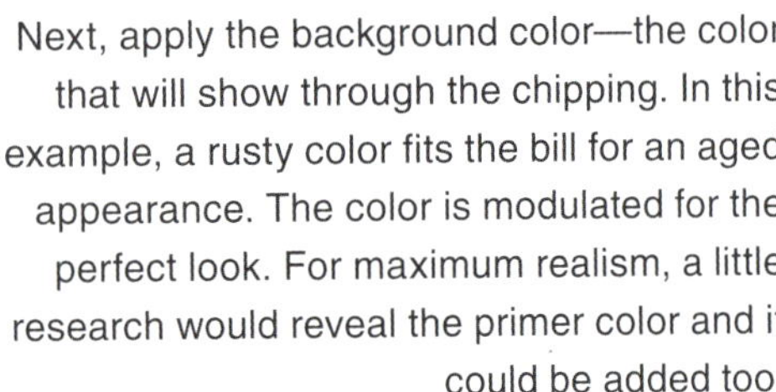

Next, apply the background color—the color that will show through the chipping. In this example, a rusty color fits the bill for an aged appearance. The color is modulated for the perfect look. For maximum realism, a little research would reveal the primer color and it could be added too.

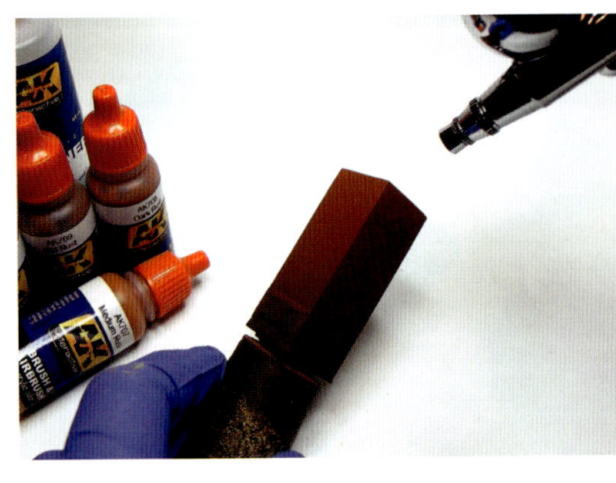

Once the base coat has dried, brush water onto the areas you want the rust to appear. Once moistened, cover the area with salt. The size of the grains will determine the size of the paint chips. If your model has several sides, wait for each to dry before proceeding to the next. If too much salt is spattered on the surface, remove it with an old brush or a toothpick. If you prefer, you can use pigment fixer instead of water. The result will be exactly the same.

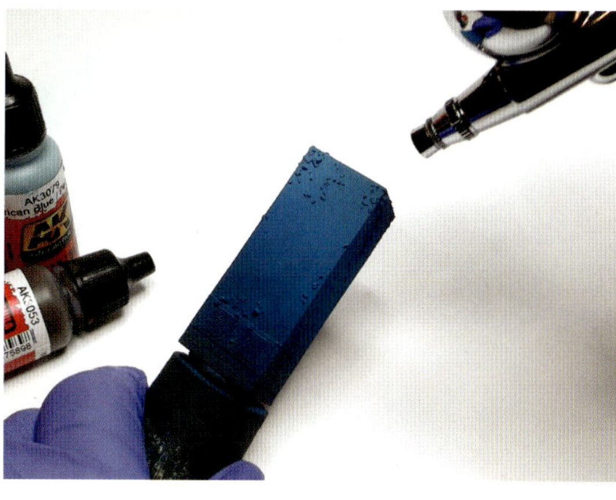

After the salt has set, paint the surface color. Finish as normal, complete with highlights, shades, and color modulation. Although you can brush-paint the surface, an airbrush will provide a more realistic finish.

When the surface paint is dry, remove the salt with a toothpick or an old brush with stiff bristles. Take care not to scratch and damage the surrounding paint areas. You can further enhance the chipping by painting a light-colored outline around the chips. This provides a more dimensional and eye-catching result. Also, to improve realism, you can create some scratches, making random lines of black or rust color. As with the chips, illuminate them by adding a light-colored outline around them for maximum effect.

The paintwork may end here, or be combined with more weathering techniques. Either way, apply a coat of varnish to protect the work so far. While the vending machine received a gloss coat, use a varnish appropriate to your model and what you intend to do next. Before applying this technique, thoroughly research your subject. That way you'll know what colors to use to accurately represent the weathering and damage.

TIPS

- Salt is water soluble. Too much water on your model may dissolve the salt into your paint leaving white residue marks that are difficult to remove.
- Use salt of different grain sizes to create irregular and random chipping.
- When selecting the grain size, always consider the scale of your project.

Soap technique

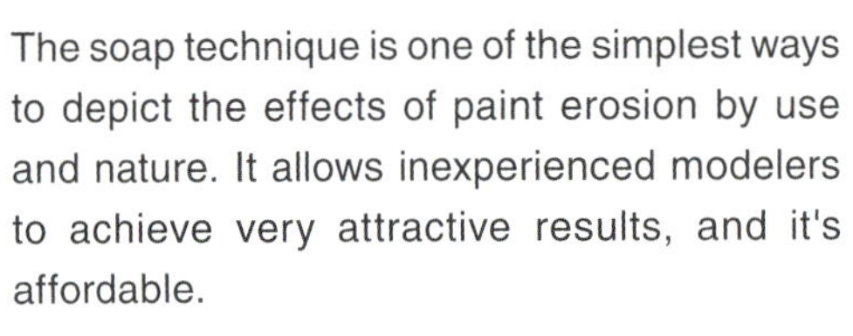

The soap technique is one of the simplest ways to depict the effects of paint erosion by use and nature. It allows inexperienced modelers to achieve very attractive results, and it's affordable.

You will need a container large enough to comfortably fit your model submerged in a mixture of water and powdered laundry soap.

METHOD

Airbrush the color you want to show through the worn paint. It can be rust, metal, or any other color, depending on your subject. As before, do everything you normally would to finish, including color modulation. All painting must be done either with enamels or lacquers. If it isn't, apply an enamel varnish to protect the underlying paint so the soap doesn't wash your work away.

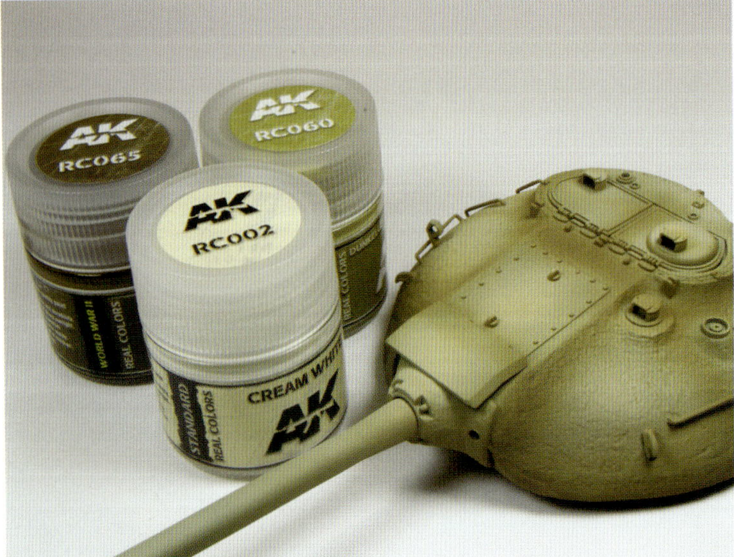

At this point, paint your model with the selected acrylic color.

Again, also create highlights and shadows on this layer, too. Don't wait until the paint is completely dry; it's enough to be dry to the touch.

Completely submerge your model in the container. Take the model out of the water and then sprinkle the soap on those areas where you want to apply the effect. The soap will stick to the wet surface.

Wait about 2 minutes so the soap sticks well to the paint.

Submerge the model again in the container with the water to activate the soap and make it dissolve. Gently shake the model in the water. As the soap dissolves, it will take the paint with it.

Leave your model to thoroughly air dry. Once dry, airbrush on a coat of varnish to seal and protect your work.

TIPS

- Remember the initial layer has to be painted with enamels or protected with an enamel varnish, otherwise the soap will also remove the base paint.
- You can use a hair dryer set on cool to dry the model. Remember, heat will warp plastic and resin.
- This technique wonderfully replicates desert wear on a paint scheme.

Streaking grime and deposits

Dirt. You can't escape it. And over time, it accumulates on everything. When wet, dirt can cause deposits or streaking, both of which are interesting to model.

Deposits occur when a dirty liquid falls on a horizontal surface without a slope or anywhere to go and ends up drying, leaving a stain or crust. This kind of dirt often accumulates in recesses and crevices.

Streaks, also known as **streaking** among modelers, are created when a liquid that carries dirt flows down a vertical or angled surface, leaving a mark in its path. The length of this mark depends on the surface. One of the goals of the weathering process is to re-create these two types of dirt. Although one could try to do it using diluted pigments or paint, there are currently many products made to reproduce this effect.

Moreover, they are made to represent certain staining agents like oil, rust, and mud.

Be very careful when applying weathering effects. For example, we can't weather a winter-camouflaged vehicle with dirt produced by desert dust. Try to be true to reality and study the environment in which our subject was deployed, works in, or is now sitting derelict. Below, you'll see how these streaking and deposits work on the remains of an abandoned vehicle. As this technique is complementary to others, it can and should be used with other weathering, including chipping and rust. To protect all previous effects, always spray on a protective layer of varnish.

DEPOSITS

To simulate dust or rust deposits, you follow the same procedures. Using enamel-based products specifically designed for these effects, you'll need mineral spirits (white spirit) as a solvent. You'll also need two synthetic brushes: one fine (No. 4) to apply the deposits with accuracy, and another thicker brush (No. 6) to blend the effect.

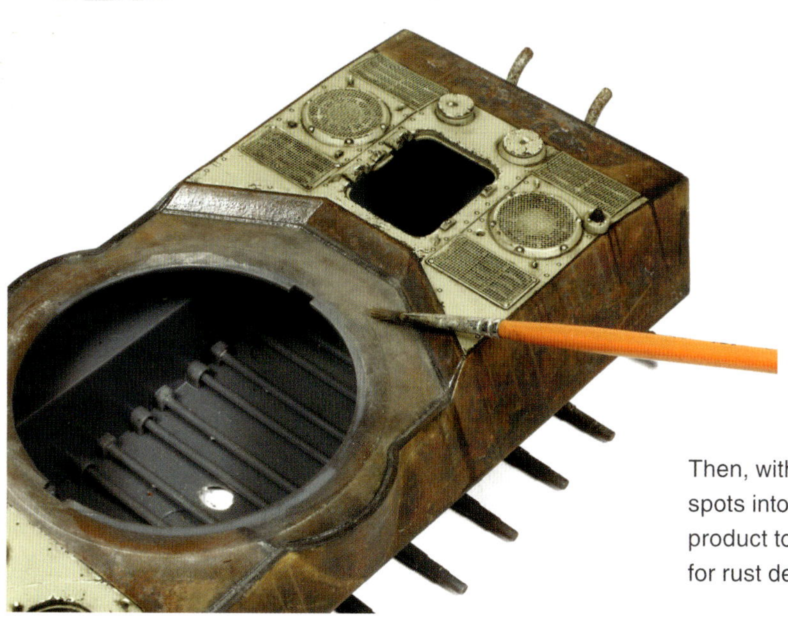

This kind of dirt of is created when a liquid containing dirt rests on an area until it dries to a crust. Select those areas most prone to accumulation, and apply the medium with the brush straight from the bottle.

Then, with a brush moistened in mineral spirits, blend these spots into your paintwork. Wait a couple of hours for the product to dry, and then check the final result. Do the same for rust deposits.

RUST AND STREAKING GRIME

This is a much more obvious effect than the previous one, but it is also a bit more complicated to pull off. Again, streaking grime forms when a liquid carrying dirt flows down a slope or vertical surface leaving a trail in its path. To simulate rain streaks sliding down the side of a train car, we'll need the streaking medium, mineral spirits, and two brushes as before.

Apply this effect only to vertical or sloped surfaces. Draw vertical lines with the medium, using a fine brush. Each streak starts from a rust or dirt mark, and flows downward. The trail should be longer on surfaces with a steeper slope. The thickness of the streak depends on the size of the starting mark. In the case of rain-streaks, the lines should be thicker at the bottom; due to gravity, dirt will accumulate at the lower end.

With a flat brush slightly moistened in mineral spirits, blend the lines, pulling the brush from top to bottom, until you get the desired effect.

TIP

- You can easily remove these products if you don't like the results. Simply wipe them away with a brush or cotton swap moistened with mineral spirits.

How to make textures and mud

Dirt, dust, and mud are elements that are to be found on virtually any vehicle operating outdoors.

Even a brand new Formula 1 car carries a lot of dirt from the asphalt after a race, even if the circuit is dry. On tanks or military vehicles, this effect is more intense. Airplanes have a lot of grime on the landing gear, the interior, and on areas close to the propellers. There are currently many different products available to make dirt textures. You can find texture products with different color finishes, such as dry mud, wet mud, and even mixed with fibers to imitate pieces of vegetation. Of course you can always create your own plaster mixed with different earth and vegetation elements.

One of the most important aspects when trying to represent textures is to choose a color relevant and appropriate to your model. For example, you can't use an orange mud-tone from the iron-rich earth of Vietnam on a vehicle operating in Europe during World War II. Yes, even mud must be correct to preserve the historical background of our work.

Admittedly, there is more freedom when it comes to fantasy and science-fiction models. Some modelers use mud to cover assembly or paint faults on their models. Don't let this become a habit.

The technique of splashing mud with a brush is very useful for replicating a thick accumulation on the undercarriage of a vehicle. You can also do it by blowing air from an airbrush through a paintbrush. Increasing the pressure, the splatter of mud will vary, giving you different accumulation effects.

Mud spatter created by tapping a paintbrush.

Flicking mud using a toothpick.

Mud spatter made by blowing air from an airbrush through brush bristles.

Simulating mud is relatively easy. Still, you need to have a clear idea of the amount and position to make it look realistic. The mixture to get a mud or earth texture will vary depending on what you want to depict. For dry mud, you can make a mix of light colors with pigments, enamels, or both. For wet mud, you can use a mixture of darker colors, adding a few drops of varnish or wet effects.

The lower areas of the chassis on tanks or vehicles should be effectively weathered. Concentrate more on the wheels so the accumulation looks more realistic.

It's not necessary for a texture to be very thick. Sometimes a thin mixture can give great accumulations of dust and very realistic textures.

The technique of adding textures not only involves making mud, but also imitating effects such as rust. Textures are not limited to land vehicles and figures. They can also be applied on dioramas and aircraft.

Rusted texture with sand

If you carefully examine a surface with heavy oxidation, you'll notice that besides changing color, it also generates a rough texture, which chips the paint and makes it flake off.

You don't need to create this texture in small scales, since it would look oversized and unnatural even with the finest grain. Paintwork is enough to simulate the rusted surface. However, when working in larger scales, you need to find ways to simulate that texture for maximum realism.

Most models are not designed with different textures, and certainly not with rust textures. To replicate such textures you'll need, beyond the correct colors, some fine grain sand and a fixer.

You have to be careful choosing sand. The grain can't be very course because it will look out-of-scale. Yet, not too fine either, because the desired effect will be too faint.

The fixer we're going to use for this technique must be fast drying and have good adhesive qualities, bonding sand to plastic. Super glue fits the bill.

Before you begin gluing the sand to your model, airbrush on a coat of primer.

1. Decide which areas will be rusty, then apply the super glue.
2. Next, sprinkle the sand onto the areas covered with glue.

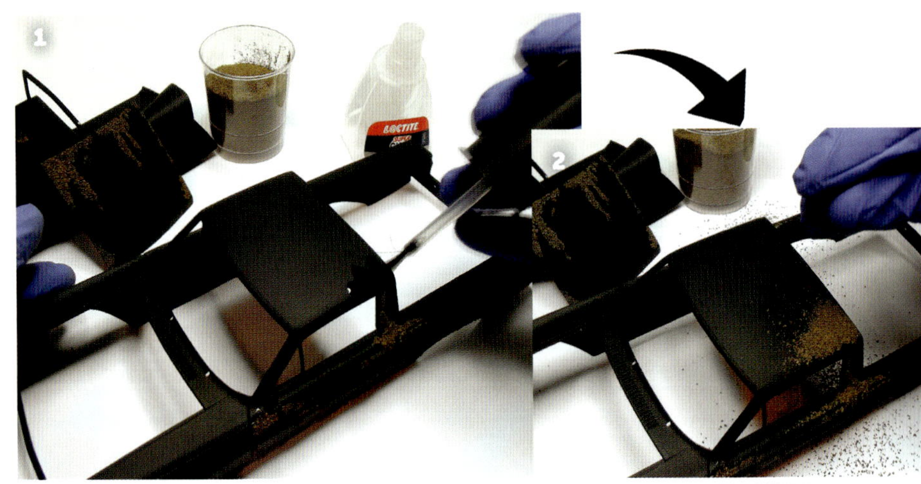

3. After the glue has set, gently shake your model to remove loose and excess sand. Follow up by brushing the areas with a soft paintbrush to remove any remaining loose grains.

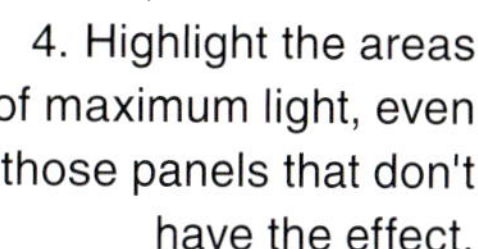

4. Highlight the areas of maximum light, even those panels that don't have the effect.

5. Airbrush the textured areas your rust color. Don't worry about staining the surfaces around the texture. Real rust would stain them, too.

6. Use the sponge technique to further enhance the appearance of rust with reddish ochre tones.

At this point, airbrush a coat of flat varnish on your work to protect it. The following steps are quite aggressive and could damage the surface.

7. Brush chipping medium on all those areas where you want to see the rust effect.

8. Cover the whole model with very thin and transparent layers of the acrylic top coat color (light green in this case) so the previous highlight work is visible.

9. To finish the process, soak the areas with water to activate the chipping medium. With the help of a stiff brush, also moistened in water, remove the color from the textured areas. Then, airbrush another coat of varnish again, to seal and protect your work.

TIPS

- When adding the sand to your model, work in small areas in order to save material and time.
- Your painting is not necessarily the end to this process. Use other weathering techniques on it.